"In the US inventor community, there are only a handful of experts whom I trust to be honest and provide high-level knowledge to support my clients. Warren is one of those experts. Twenty-five years in the industry and a heart for protecting inventors from scammers—I can vouch for Warren without reservation."

—Kedma Ough,
Small Business Super Hero and Author of *Target Funding*

"Warren is one of the primary people responsible for the Inventor Spotlight Pavilion at the National Hardware Show growing into the largest inventor booth event in the country. Between product-pitch panels, speakers, and educational sessions, this area has become a vibrant showcase feature for our entire show."

—Rich Russo,
Director at Reed Expo and the National Hardware Show

"Warren has been a terrific friend over the years and has traveled to Chicago many times on his own dime to speak to our nationally renowned inventor organization. He always seems to have time to help everyone, without charging!"

—Calvin Flowers,
President of the Chicago Inventors Organization

"Warren is the most resolute advocate for inventors and champion of innovation I have known in my career. His knowledge of intellectual property protocol is extensive, particularly when it comes to the hurdles faced by the independent inventor community. His participation for many years on the Pro Bono Advisory Council has been invaluable to ensuring that patent pro-bono programs truly meet the needs of their constituents. I am grateful to call him a friend and be able to work alongside him in the push for innovation in America."

—Jim Patterson,
Patent Attorney and Founding Partner
of Patterson Thuene of Minneapolis and
National Patent Pro Bono Advisory Council Chairman

"As a member of the United Inventors Association Board of Directors, I have worked with Warren on several terrific USPTO events, and I've witnessed his passion for providing invaluable and credible resources to inventors. As a patent attorney and inventor advocate, it is encouraging to see Warren share his knowledge about product development, licensing, and intellectual property with the inventor community. Warren has a keen eye for recognizing the 'next big thing'! I admire his dedication to the independent innovation community." —Andrea Hence Evans,
Patent Attorney, Owner of EvansIPLaw,
Former US Patent and Trademark Office Examiner, and Author

"Warren and I have spoken often together at educational inventor events around the country, as well as participated on many new product-pitch panels. He has put a ridiculous amount of time and resources into the United Inventors Association. I know this because I have served on the UIA Board for many years. No one I know is more involved in the well-being of the US inventor community." —Scott Hynd,
QVC On-Air Product Expert
with more than $1 billion in TV sales worldwide

"What a gift this book is for inventors and innovators of all types! Warren has given us a comprehensive, highly readable, and entertaining guide to the world of invention, showing the full range of opportunities and pitfalls for the people creating the next economy." —Bo Burlingham,
Author of *Small Giants* and *Finish Big*

INVENTOR CONFIDENTIAL

THE HONEST GUIDE TO PROFITABLE INVENTING

WARREN TUTTLE

WITH JEFFREY A. MANGUS

An Imprint of HarperCollins

Published by HarperCollins Leadership, an imprint of HarperCollins Focus LLC.

ISBN 978-1-4002-1958-2 (eBook)
ISBN 978-1-4002-1957-5 (PBK)

Library of Congress Control Number: 2020951029

Printed in the United States of America
20 21 22 23 LSC 10 9 8 7 6 5 4 3 2 1

I dedicate this book to two people.

The first is my wife, Lynn, who has always been there for me, good days and tough ones, for better or for worse. Your quiet determination, tireless work ethic, and life accomplishments are truly inspirational. Thank you for your never-ending inspiration, unconditional love, and support.

The other is Bob Reichenbach, boss, mentor, and friend, who used to motivate me by saying: "Look not at how things are, but as they could be."

Disclaimer

Throughout this book, including the title, I use the traditional term *inventor* often. Feel free to substitute maker, product developer, and designer, because this book is also dedicated to you. I specifically chose not to use the much broader term "innovator" more often, as my primary focus is on independent creators.

Contents

SECTION TWO
Inventors "Beware"

SECTION THREE
Inventors "Be . . . Aware"

SECTION FOUR
Organic Innovation Matters

Foreword

By Frank Brown

Warren and I have been lifelong friends. We have shared some amazing experiences together including traveling throughout the world; attending plays, concerts and sporting events; and watching each other's children grow up into wonderful young adults. I can attest firsthand to Warren's zest for life, which is both remarkable and inspiring. Even more telling is his determination, in both personal and business pursuits, to make things happen. Warren is more than talk, he's a doer. He's started nine businesses and managed more than a half-dozen non-profit missions. While some may think about planning national inventor events, speaking engagements, and new product pitch panels, Warren turns them into reality. Others may imagine putting a golf trip for sixteen to Ireland together, he actually makes it happen, and rather seamlessly I might add. The list goes on and on.

Warren's penchant for innovation is based upon real-life experience and hands-on leadership in the field. I have spent many early morning walks listening to Warren carry on about what America needs to do to maintain its place as the startup capital of the world. He eats, sleeps, and breathes the subject. He's a tireless advocate for grassroots inventors, product developers, and makers, who together make the United States economy truly unique, continuing a historic tradition since our nation was born in the late 1700s.

This book is a thought-provoking read for anyone who is an entrepreneur, creator, or innovator who participates in any way within our vital innovation ecosystem. It's also a book for corporations looking to profitably expand their open innovation programs and better interact with the outside world. Not all great products are born in corporate labs. Many new products come from the basements, garages, and workshops of independent innovators from around the world. It's time more companies link directly with these folks and take advantage of this true natural resource.

Warren is also an advocate for all of the traditionally underrepresented innovators in America. Besides leading the 501(c)(3) educational non-profit United Inventors Association, he is an original member of the National Patent Pro Bono Advisory Council (PBAC) which provides low-income innovators the opportunity to file patents for free. He also serves on the Department of Commerce's National Council for Expanding American Innovation (NCEAI) which is identifying ways the innovation space can be more inclusive for minorities and women. Warren is passionate about everyone having access to strong patents and a flourishing innovation ecosystem.

Told from the perspective of a life and career filled with memorable experiences, this book adds great value to the innovation landscape. Given today's rapid changes in technology and medicine, there has never been a better time in our history to take a reconciling snapshot of our innovation landscape; remembering where we came from, analyzing where we are today, identifying the real challenges facing innovators, and determining where we should be headed in the future. *Inventor Confidential* tackles these subjects head on. After all, the next generation of innovators are counting on us to keep the ship on course.

J. FRANK BROWN is Managing Director and Chief Risk Officer of global growth equity leader General Atlantic (https://www.generalatlantic.com/people/frank-brown/) and author of *The Global Business Leader: Practical Advice for Success in a Transcultural Marketplace*.

Preface

Before Getting Started . . .
Let's All Take This Inventor Pledge Together

In business, there are two paths to profitability: generating revenue or reducing expense. As an inventor, think of profit as more money in your pocket, and not, as typically happens in our industry, more money in everyone else's pockets. Here's another way to think of it: If you went to great effort and expense to bake a pie, why would you give away every slice and leave nothing for yourself? If the answer is that you baked the pie solely for creative satisfaction and a heartfelt desire to feed others, this book may not be for you. In the invention arena, there are plenty of people on all sides who will consume every single slice of your pie without any concern for your financial well-being. They may talk a good game, but the results are indisputable.

I will focus on both paths to profitability in this book because not enough inventors treat their new product pursuits as cost-effective, profit-oriented businesses. More often, inventors approach their craft as a hobby or game. This leads to being taken advantage of, including in many ways you are probably not even aware of. There are people seeking to profit from you and your efforts, and they are counting on your naivete to remain in business.

You may think you are unique, and there is no doubt that in many ways you are, but invention industry marketers have seen you, the wide-eyed inventor, a thousand times before. They know how to reach you and leverage your insecurities. They know precisely what key marketing words to use, and which calculated stories to pitch. They project their business revenues years in advance, knowing that you are coming. These marketers count on would-be inventors rolling in and out like the tide, maintaining their own income stream, while your financial success as their client is usually zero. They leverage your dreams for their own profit. Behind all the clever words, their business model is constructed for their benefit. The rest is a well-orchestrated front, or in some cases, multiple fronts. There is little in the way of win-win, like most other business relationships require. It's usually just a win . . . for them.

At the same time, on the political front, big tech companies and their lobbyists resolutely attack our historically successful patent system, as they try to rewrite hundreds of years of intellectual property law to benefit their own market-dominant positions. They are well-funded, organized, coordinated, connected, and highly efficient. They recognize the inventor community is under-funded, unorganized, naive, and, as a collective interest group, highly dysfunctional. Efficiently infringing on others' intellectual property these days has become a financial risk worth taking. They unduly influence Congress and the nation's courts and do not want inventors sitting at the national innovation discussion table, even though some of these same big tech company leaders started their careers as grassroots innovators themselves. Much like invention industry marketers, they do not have your interests at heart, nor do they have much concern for the future of organic innovation in this country. They are myopically focused on their own well-being, profit, and future. Guess what? You should be too.

Let's proclaim a new day for inventors and all take this pledge together.

THE INVENTOR'S PLEDGE

INVENTORS, REPEAT THIS PLEDGE AFTER ME:

"I _______ have a right to either make a profit or protect myself from financial loss. I will stop thinking naively and position my own fiscal interests ahead of those invention industry marketers who would profit from my creative efforts without generating more revenue for me than the fees they charge. I also need to become more aware of those industries, institutions, and politicians that would rescind our historically protected property rights, provided for in the US Constitution Article I, Section 8, which states: 'Congress shall have the power to promote the progress of science and useful arts, by securing for limited times to authors and inventors the exclusive right to their respective writings and discoveries.'"

Do we all agree to this pledge?

If not, skip the rest of this book.

If so, then let's get started.

Independent inventors, one of the bedrock treasures of our society, and historically the backbone of American innovation, commerce, and job creation, are under assault from all sides today. Flashy marketers cleverly take their money while very rarely providing any back-end financial success. Large companies lobby in Washington, DC, to weaken the value of patents, which makes infringement on independent inventors easier. The Supreme Court is changing the Constitutional definition of patents from the historically accepted twenty-year protection of actual property rights to government franchises that can be rescinded. Congress passed the misnamed Leahy-Smith America Invents Act (AIA) in 2012, a compromise between large pharma and tech companies supposedly aimed at curbing patent troll abuses—yet adversely affecting independent inventors in the process. In fact, as crazy as it sounds, the bill included a provision that allows the same federal organization that carefully reviews all intellectual property applications and grants patents after years of review, the US Patent and Trademark Office, to relatively easily rescind those issued patents, much like an umpire changing the score of a baseball game after it's over.

I have been immersed on the front lines of this brave new innovation world for the past several decades, bringing disruptive products to market, serving multiple companies as an Open Innovation Director tasked with securing profitable outside licensing agreements, and leading a national non-profit inventor organization, which, among other educational pursuits, helps advocate for inventors in Washington, DC. I have observed significant changes during this time, quite possibly more change than at any other time in our country's history. Though, at first, many political tweaks appeared subtle, taken collectively, they are remarkable.

Alert: We are well on our way to large corporations stifling future grassroots initiatives and organic innovation.

Most Americans are aware of today's severe environmental and fiscal challenges. Many are not aware, however, of what's happening on the innovation front. It's not a very sexy subject . . . yet. Without a protective period for incubation (which was well considered and addressed by our founding fathers in 1792) providing independent profit motive, creative and organic innovation will dramatically recede in our country. Without incentive, grassroots inventors will drop out of the game, and with that, most of the new products and services in our future will come only from large companies. We know what this means for America in the long term: more and more risk-averse decisions from sheltered boardrooms. Companies that have achieved market dominance will inevitably cut costs on new product development. Innovation becomes vanilla. Think of the auto industry in the 1970s and 1980s. Thank God for the competition provided by foreign car companies, or we'd all still be driving Corvairs. Or think of Kodak inventing the digital camera and then shelving it because somebody at the top determined it would disrupt their film margins.

So, what can independent inventors do?

Let's start by shining a light on the many ongoing challenges we face and, with some honest awareness, begin by righting the ship. Let's recapture the spirit of our country's historically great inventors. Then, let's share inside information on how to best approach the marketplace to either increase your odds for financial success or, if necessary, pull back and save your money while living to fight another day.

If you are an inventor, product developer, designer, patent holder, maker, or even a dreamer, my primary goal in this book is to expand your understanding of taking a product to market while saving your time, energy, frustration, and, most of all, your money. There are many inventor books available written by inventors who have experienced varying degrees of marketplace success. These books typically include step-by-step, "how-to" informational guides, featuring a variety of getting started checklists. If you are new to product development and the world of inventing, some are worth reading. It's always a good idea to gain valuable insight from other knowledgeable inventors.

Correspondingly, if you are associated with a larger company, I will be advancing the case for creating open innovation programs that ethically and profitably engage corporations with the outside world. I will explain how to accomplish this, as I have been in the trenches for a long time and have learned a great deal. Innovation will serve as a prime conduit in digging out of our nation's latest financial setbacks. Not all great ideas come from within your company. It may be time for you to look for new blue oceans to add profit.

There are also a multitude of inventor books written by middleman marketers. Assuming you sign up for their associated programs, they claim to share the path to gold mines, even millions of dollars. Though there is some helpful dogma offered in these books, most are promoted with blatant and subliminal messaging that getting rich is just around the corner. This financial lure is irresistible for many newcomers who don't know how to get started. The untold truth is, such authors make far more money selling their books and adjunct support programs than inventors earn following their get-rich advice. The fact is, making money inventing is neither simple nor easy. If it were, they'd be doing it. The sooner you realize this, the better off you will be.

This book is entirely different. I am a practicing, full-service, hands-on open innovation expert. I am the real insider, with both experience and repeated success. I am not a marketer. I am not a reporter. I am not interviewing others who may have varying degrees of thoughts on the open innovation subject. I eat, sleep, and live 24/7 in the product licensing arena and make decisions about disruptive products every day. I also volunteer a considerable amount of time supporting inventor issues, organizations, and councils; travel regularly to Washington, DC, for various causes; and am tied closely into the national inventor club community.

This book is about a lot more than simple ideas. Most large companies featuring open innovation platforms are not looking for your ideas, which are a dime a dozen. Everybody has them. I skim through them every day without taking many seriously. Highly successful companies that repeatedly license outside inventions are looking for proven, innovative technology that can be exclusive for them, ready to advance and fit into their plans, and worth paying for. This book is about serious product development, filed intellectual property, skillful negotiation, and licensing with reputable companies. These are companies that have manufacturing facilities, fulfillment capabilities, identifiable brands, access to retail shelf space, consumer trust, and a proven track record of new product sales.

I am not trying to sell inventors anything. Instead, my goal is to explain what the view is of inventors from the inside of companies and make you aware of all the serious challenges involved in getting your product to market before spending your hard-earned money and ending up with nothing but debt. As for the other books written by middlemen-marketers who charge inventors up front, I say: Show us your numbers—the real numbers. How many inventors have you taken money from? How much did they pay you? How many total licensing agreements were generated? How many inventors earned more money in sales royalties than they paid you in fees? And what is the percentage of total royalties collected from inventors that you contracted with compared to the fees you charged them? Pretty easy questions. What are your numbers? (Quite honestly, I would also want them audited. You know, trust but verify.)

Though I will share specific inventor stories and marketing lessons I have learned along the way, this book is written from my extensive experience working on the inside of companies. I have been responsible for many profitable product launches over the past twenty years. I want to share with inventors around the world how the licensing process really works from the inside, the real story that has not been fully told. Over the past fifteen years, I have directly connected with well over 100,000 inventors via product submissions, phone calls, emails, texts, social media posts, trade shows, industry conferences, and speaking engagements, while absorbing a great deal of knowledge along the way. The creation of personal wealth and maintaining cash flow have never been the primary drivers of my business or non-profit decisions. I do not charge inventors, as my services are paid for by the companies I contract with. This model allows me to pursue what I love most in life, helping other people, even when it includes advising them to save their money and resources, or to throw in the towel.

HONEST TIPS

Inventors Digest Millionnaire Poll

The percentage of inventors who think their invention will make at least $1 million, according to a blog at www.edisonnation.com: a whopping 78 percent! Just as astonishing, 54 percent of inventors think their invention will earn them over $5 million. Here's a reality check: Fewer than 3 percent of inventors sign licensing agreements. A fraction of 1 percent of inventions earn over $1 million for the patent holder.[1]

I dedicate this book to those who have a big idea and want to explore taking it to market the right way. I will help guide you in setting realistic expectations learned from substantial hands-on experience. I will explain why signing a licensing agreement is only the first of many steps to making money and not an endgame achievement in and of itself. I will prepare you on how and where to pitch your products, as I pull back the curtain on the inner workings of corporate innovation programs. I will provide a firsthand view of real-world discussions and product evaluations from inside the decision-making boardrooms. Most of all, I will lead you down an alternative path and help you succeed by pursuing an honest approach to innovation.

But wait, there's more!

Inventors today face other significant challenges. Intellectual property laws have changed on Capitol Hill. A Patent Trial and Review Board (PTAB) has been established at the United States Patent and Trademark Office that allows for review of patents even after they are issued. And the Supreme Court has made some highly questionable rulings of late on what determines patentable material. The courts have begun to drift away from the constitutional definition of what a patent actually is, changing it from a property right to government franchise. It seems like many years ago, independent inventors were celebrated and highly valued in this country. When it comes to innovation today, however, we tend to over-celebrate the five largest tech companies: Facebook, Apple, Amazon, Microsoft, and Google. Ironically, these companies were all started by independent innovators. Yet, they have rapidly grown into huge corporations now doing their best—through massive lobbying efforts—to rewrite the very same protected intellectual property laws that enabled them to get off the ground only a few short years ago. They are behaving like little kids playing in the treehouse: Once safely up, they are pulling up the ladder to exclude others. Some feel quite comfortable raiding the intellectual property of others. It's becoming far more cost-effective in recent years to willfully infringe because of the significant legislative efforts these companies have unleashed upon Washington, DC. We'll take a closer look at how these challenges affect inventors.

Remember, throughout the book, I'm an end-user and a decision-maker, not a middleman. I am a buyer, not a seller. There is a big difference. I am the person the middlemen come to when they want technical advice or have a product in my field they wish to license. I run successful open innovation programs and review many thousands of new product submissions each year. I work closely with senior corporate managers to select new products and advance the winners. What I do is unique, and I want to share the inside game with you.

This book is not about mailing a product idea in and hoping for the best. It's not about sketching a product, sitting back, and collecting checks. It's not about reducing your workweek or about product development being simple. This book is not about gold mines and hitting the lottery. It's about the hard work and dedication necessary to do things right. This book is about learning how to leverage inside knowledge and become aware of all the challenges that surround you today in order to gain the best odds for your own personal financial success.

From time to time, I have this recurring vision of inventors as Hobbits: J. R. R. Tolkien's wonderful, well-intentioned farmers living idyllically in the Shire while, unbeknownst to them, the rest of the world is at war. They surely don't stand a chance against all the outside evil forces who stand twice their size and strength. Yet, guess what . . . the Hobbits have responsibility for the ring! And when it comes to innovation, so do inventors. As we look to the future, let us never forget this. I think we all need to start thinking again, just like they did back in 1792, of ways to make the world of organic innovation more productive by tapping into, and rewarding, the motivation and potential of America's independent inventors.

Introduction

Living an Innovative Life

Who am I exactly, and what provides me the platform to write this book? Here's another way of asking the same question, taken from the pages from Malcolm Gladwell's book *Outliers*. Where have my 10,000 hours of experience come from?

I have always lived an out-of-the-box, innovative life. My name is Warren Tuttle, and I'm a well-recognized, hands-on open innovation expert and inventor community philanthropist. Open innovation is the art and science of companies searching outside their four walls for breakthrough products and platforms they can manufacture, market, and sell under their proprietary brands. I do not have a PhD. I don't utilize a large team of people. My approach to open innovation is unique, organic, and transparent. I am not as much a philosopher as I am a worker. I liaison inventors, makers, designers, product developers, and their disruptive products every day with several industry-leading consumer goods companies.

Then, I help integrate those products into the marketplace through fair and productive licensing agreements that benefit both innovators and companies.

I currently oversee, or have overseen in the past, the external product search and licensing programs for three corporations: publicly traded housewares and tabletop icon Lifetime Brands, top-selling hardware industry pacesetter Techtronic Industries Power Tool Group, and direct response television stalwart Merchant Media. I also serve as president of the national 501(c)(3) non-profit United Inventors Association, dedicated to inventor education and advocacy. How I arrived here has been one heck of a ride. I am at ground zero of the United States innovation arena and have had the privilege to work with many incredible people while helping grassroots innovators pursue their dreams. I love what I do, and the best part is, I do it without charging inventors, which, for several reasons, is important to me.

A terrific book that has influenced my open innovation pursuits is *Blue Ocean Strategy*: Written by preeminent management thinkers W. Chan Kim and Renee Mauborgne of the world-renowned INSEAD Business School based in Fontainebleau France, *Blue Ocean Strategy* recommends that businesses rethink new approaches to traditional growth and find out-of-the-box, blue ocean arenas to sail into and pursue, instead of spilling red blood fighting their competition every day to maintain market share in tired old seas. Jeff and Dan Siegel of Lifetime Brands had just read *Blue Ocean Strategy* when we met to establish our novel open innovation program there. The book had a tremendous effect on our launch, and what better way is there to pursue new arenas than looking outside the company for disruptive products and platforms?

I believe that everything I have done in my life has prepared me in some critical way to develop the open innovation programs and non-profit inventor organizations I oversee today. My personal life story is a litany of unforeseen disruptions and the willingness to adjust and pursue unique life choices. At the age of twelve, my father, who spent a career working for mega-corporation General Electric, announced to our Connecticut-based family that we were being transferred to Zurich, Switzerland. This was my first significant life disruption. Then, at the ripe age of sixteen, just as I was getting used to life overseas, I moved many thousands of miles back to the United States on my own to attend prep school and, eventually, college. I have lived on three continents (North America, Europe, and Africa), traveled extensively around the world, and even as a youth hitchhiked coast to coast to experience the entire United States firsthand, sometimes sleeping on the side of roads. I have climbed Mt. Kilimanjaro, run with the bulls in Pamplona, helicopter skied many times in western Canada, completed the New York City Marathon, golfed hundreds of times in Ireland, and motorcycled through the Austrian Alps and Italian Dolomites. I played lots of different varsity sports growing up as the third of four highly competitive boys, including walking onto a top-ranked college lacrosse team. Although I barely missed Woodstock (thanks, Mom, I was only thirteen), I have attended many concerts since and still do.

I have also launched nine of my own businesses. I have made pretty much every mistake in life at least once, sometimes twice, some which involved very tough lessons. And yet, here I am, still standing! Flourishing, in fact. Rain or shine, my life has been unapologetically dedicated to new experiences, risk-taking, the pursuit of new challenges and disruptive change, and the unabashed acceptance of those different from me. I'm in the "there is always a better way to do things" camp. Plus, I love people. All people. Especially those different than me. I think that's why I feel so comfortable in America's grassroots innovation space.

HONEST TIPS

Third Culture Kids (TCK)

Third culture kids are children raised in a culture other than where they were born. TCK are often exposed to a greater variety of influences as they move between cultures. There are benefits to being a TCK. They usually understand there is more than one way to look at situations, and, with an increased number of hands-on experiences in multiple cultures, there is a difference in how they perceive the world. On the other hand, TCKs can experience challenges returning to their home culture as their expanded worldview is often seen as useless to their monoculture peers.

I have been a hard worker all my life. I had a paper route when I was ten, which taught me at an early age to focus on business and manage my finances. As I got older, I shoveled sidewalks and plowed driveways during snowstorms. I started a successful window washing and painting business during my college years. These activities taught me that I could run profitable endeavors. I also washed dishes and bussed tables at night for a local restaurant, a terrific lesson in how to do the same monotonous tasks over and over for hours on end without getting discouraged or lazy. I recognized from an early age that there is no substitute for working hard and mentally striving to do a great job, no matter how menial.

Merchandise Buyer in New York City

Before graduating from college, I got serious about starting my full-time career and discovered a department store chain called Abraham & Straus. Their flagship building was in Brooklyn, NY. It was founded in 1865 and became part of Federated Department Stores. Today it has merged into Macy's. A&S was also a sister store to Bloomingdales. The main store was a massive building with over 850,000 square feet, the third biggest freestanding store in the entire world. What intrigued me was Abraham & Straus had a tremendous merchandise buyer training program—at the time the best in the country. It represented an excellent opportunity for a young man fresh out of college to start a career. I always wanted to be my own boss, so I was sold on managing my own department as if it were my business.

I graduated from college on Saturday, May 5, 1977, packed my belongings, and started working that Monday morning, May 7. I began in the merchandise buyer training program with 105 recruits in the class. I'll never forget when the company manager gathered us and said, "Welcome to Abraham & Straus. Take a good look around because, by the end of the summer, you won't see too many remaining faces."

I was one of the only kids from Connecticut. Most of the kids sitting in that room were from New York City, and a lot of the trainees came from families with generations spent in retailing. Abraham & Straus had stores all around the New York area, with branch locations in Queens, Manhattan, Long Island, Westchester County, and New Jersey. From my class, only five of us eventually became buyers. I was initially assigned to the luggage department for my first five weeks and then moved onto the tabletop area, where I spent another five weeks. Later, I was promoted to an assistant buyer in the Christmas Shop. It was a big deal; Christmas was, and continues to be, a massive holiday in retailing with a lot of products and moving parts. Next, I was promoted to the assistant buyer of the cookware department. Cookware was one of the biggest and busiest departments in the store, and thus began my lifelong love affair with the housewares industry. I learned how to identify and select products, how to track shipments and sales, and how to negotiate deals with suppliers and companies all over the world. I honed my skills while my confidence grew. I became a profit margin expert. I learned about fulfillment, marketing, and product display, all on someone else's dime!

Then one day, my boss and mentor, Bob Reichenbach, promoted me to the buyer of the small appliance department, which at the time was huge for me. It was the single biggest department in the entire housewares area, and I was still only twenty-three years old. Bob was the merchandise manager and had several buyers reporting to him. He was tough and pushed me in so many ways. In my mind, he was the Bobby Knight of merchandising. Underneath, he cared about me and helped mold me into the successful businessperson I eventually became. Throughout my life, I have had to work hard to keep up. There was never any magic pill or potion. I wanted to learn everything, so I had to dig in and hone my craft from the inside out, including mastering the warehouse and how to get goods out on the selling floor more quickly. I walked many trade shows and was always willing to take a chance on a new product. That's what I loved the most, finding and launching new products.

Becoming My Own Boss

After leaving Abraham & Straus, I started and acquired a collection of retail stores and businesses in southern Connecticut. I grew the operation to six specialty kitchenware store locations, a gourmet takeout food store, a cooking school, and a kitchen design business. At its height, my company had sixty-five full- and part-time employees, with a million-dollar payroll, which, for suburban locations at the time, was sizable. The stores were named The Complete Kitchen and The Good Food Store. *Food & Wine Magazine* once described The Complete Kitchen as "one of the finest gourmet stores in America." My culinary hero Julia Child, whom I met many times, visited our store for a book signing, and we had many other leading chefs from around the world teach cooking classes. Long before the Food Network, we were the place to meet celebrity chefs. On the food side of the business, one of my business partners was the iconic actor Paul Newman, about the same time he launched his own non-profit gourmet food business, Newman's Own. The company received continual buzz, and I loved running around like a crazy man. I had to oversee a diverse group of employees, meet payroll, lease equipment, pay rent, create marketing flyers, manage fulfillment, and so much more. I also served as president of the local Chamber of Commerce and, for eleven years in my spare time, oversaw a historic downtown revitalization program designed to get customers back from the malls and shopping locally. We were truly twenty years ahead of our time. Then one day, out of the blue, a gentleman walked into the store, sending my life in a new direction that I never saw coming.

HONEST TIPS

Launching Nespresso

One of the benefits of owning The Complete Kitchen was discovering new products. We were among the first stores in America to launch the Donvier Ice Cream Maker, Mouli Electric Mincer, Burton Stove Top Grill, Nantucket Peppergun, and Rema Insulated Baking Sheet. Another unbelievable product that we were the first store in the United States to launch was the Nespresso Coffee System. Today it is a huge success, and Nestle has sold billions of coffee capsules worldwide. Back in the late 1980s, we were fortunate when the director of the new program, sent from Switzerland to get the launch started in America, moved to Greenwich, Connecticut, where one of my stores was located. We did a private invitation mailing to our customer list, arranged a special evening of demos, and sold ten machines at almost $500 each the first night. The rest was history. We sold over three hundred machines in the first year.

Helping My First Inventor

I was forty-two years old the day Tom Risch walked into The Complete Kitchen. Tom was the creator of Misto, and the first inventor I ever met. He was both a confident and humble person who showed me the prototype of a product that was brilliant. Misto was a gourmet olive oil sprayer, and I loved it. I had never worked directly with an inventor in any capacity. I had only worked with wholesalers and manufacturers. I had a keen eye for new products. I would find them, negotiate the terms for an opening order, have it shipped in, and then we'd sell them. It was a little different meeting a real inventor.

Tom and I ended up talking for hours. He was not new to inventing. He had, at an earlier time in his life, launched a successful chimney sweep business called August West, but he had never been involved in the housewares industry. Misto didn't use chemicals or spray propellants like PAM, which is not safe around high heats and stoves. Misto allowed the consumer to spray their own choice of olive oil onto their food in smaller portions, saving calories. With Misto, there was no waste because the container was reusable and wouldn't populate landfills with the disposal of more aluminum cans. In my mind, this was a home run. So, I told Tom, "I want a hundred units, and you can ship them right away, and I'd also like you to pay the freight."

Tom stopped me in my tracks. "Wait a minute," he said. "I haven't even built any of these yet . . . this is only a prototype."

We agreed, to get us started, Tom would make twelve Mistos in his garage and drive them down to us himself. We sold out of those the first day. Then Tom made twelve more, and the same thing happened. After a few days, we upped the daily delivery to twenty-four. Then to forty-eight. After that, Tom had to build an assembly line in a warehouse to keep up with demand.

Thus, my introduction to the wonderful world of inventors began, as Tom soon asked me to help him launch Misto in the national marketplace. Interestingly enough, all the time that Tom and I worked together, we never signed a formal agreement. We worked together on a handshake, old-fashioned trust, and success. I began by showing the product to various people within my circle of friends to get their opinion. To my surprise, half of them did not understand it. Why? Because they were men who did not cook. Women and culinary people got it instantly. Soon after, I convinced about twenty specialty stores in the greater New York and New England area to try selling them. The product jumped off the shelf. They couldn't keep them in stock, and I intuitively knew we were onto something big. Then we sold 10,000 units to Bloomingdales in New York City, and the rest was history. We ended up selling over 1.2 million Misto units the first year and made a considerable amount of money. It was the American dream come true. I had done many things in life and been involved in many different business projects, but I had never been through anything quite like this. We were changing the world. Helping Tom bring his product to market provided a master course in launching a grassroots invention and set me on the path to learning how to help inventors better.

Searching for a New Way to Help Inventors without Charging Them

A few years later, I helped launch the housewares megahit Smart Spin. This was my fourth product to bring to market. The second one was my product, Stir Chef, which was a colossal failure. (I'll tell you all about that one in a later chapter.) My third was a unique raised edge frying pan that allowed a home chef to flip food and quickly sauté.

Saul Palder invented Smart Spin. With his product, I pursued the direct response TV infomercial approach. Smart Spin soon became a household name, and within the first year, we sold over five million units and ten million overall. It was another great ride. Working with individuals like Tom and Saul planted a seed that I could genuinely help inventors. Soon I became more involved within the inventor community. I felt drawn to learn all I could about the grassroots innovation world and pieced together a business plan that could fit into many life-learned lessons and ethical priorities. I investigated many different inventor-help company models—I wanted to know about every nuance that made them tick—and visited many inventor clubs looking for a way to do things better. This was back in 2008 when the housing market crashed along with the economy, and things were financially tight for everyone in the country. I was determined to find a new, disruptive, helpful path to assist inventors—one where I did not have to take their hard-earned money.

INVENTOR CONFIDENTIAL

SECTION ONE

Open Innovation

1

Inside Open Innovation

Open innovation is both the art and science of companies searching outside their walls for breakthrough products and platforms they can manufacture, market, and retail under their own brands. I have helped construct three successful corporate open innovation programs. What follows in this chapter is drawn from my unique open innovation experience, and it is intended for every inventor, product developer, designer, maker, and patent holder who is considering taking their own product to market.

HONEST TIPS

Open Innovation

Open innovation is a term used to promote innovation developed outside of companies and the traditional secrecy of corporate research and product development departments. Use of the term "open innovation" was pioneered by Henry Chesbrough, adjunct professor and faculty director of the Center for Open Innovation of the Haas School of Business at the University of California, who articulated this enlightened perspective in his book, *Open Innovation: The New Imperative for Creating and Profiting from Technology*. Today, many large companies have open innovation divisions, are receiving outside submissions, and are licensing disruptive products.

My Path to Open Innovation Director

Before I established my first corporate open innovation program, I needed more experience within the inventor community, so I joined my local inventor association in Connecticut. I had a strong desire to learn everything from the ground up, so I began talking to many inventors, listening to their challenges, concerns, and issues. One of the problems I often heard about was that many for-profit invention marketing companies were collecting thousands of dollars from novices but not earning back any real money for the inventors. It wasn't long before I became involved with the national United Inventors Association (UIA) and became exposed to a much larger platform of inventors. Yet, I still heard the same complaint over and over.

About the same time, in 2008, I attended the International Housewares Show in Chicago, Illinois, where I ran into my old friends at Lifetime Brands, including my former mentor and boss at Abraham & Straus, Bob Reichenbach, along with the chairman of the company, Jeffrey Siegel. Bob was then running the cutlery division of Lifetime Brands. I learned that he and everyone at Lifetime Brands had kept a watchful eye on what I had been doing over the years. They knew all about the success of Misto and Smart Spin and asked me, "Would you be interested in helping us find new outside products for Lifetime Brands?" Lifetime Brands initially started over forty years earlier when the company founder invented and patented the kitchen counter Knife Block. Grassroots innovation was clearly in their company DNA.

Today, Lifetime Brands is a large, publicly traded consumer goods corporation that has a significant internal development and design team. But before I became involved, they were not working directly with outside inventors. All product development was done inside their walls. Inventors might send them new product ideas, but those ideas often ended up ignored. Like many large companies without an open innovation program, Lifetime Brands didn't truly understand the outside inventor world.

In 2008, the US housing market and the economy crashed, but CEO Jeff Siegel had sagely decided that Lifetime Brands would innovate their way out of the national recession! This was music to my ears. We met a few times and brainstormed on what the new outside product search and submission program would look like. Specifically, I would serve as an outside contractor, not an employee. This arrangement would provide me considerable latitude for searching for great products while developing credibility within the inventor community. During this time, the innovation director of Lifetime Brands, Dan Siegal, was reading a terrific business book called *Blue Ocean Strategy* by W. Chan Kim and Renee Mauborgne. They wanted to find a way of separating themselves from their competitors. We all put our heads together and started the Lifetime Brands Open Innovation program. The mission was on.

Different Open Innovation Models

There are many approaches to open innovation, each crafted to fit the specific needs of individual companies. Thirty years ago, most large companies developed all their products in-house and kept the secret ingredients to themselves. Today there are large consumer product companies, like Procter & Gamble, who have hundreds of people assigned to their open innovation departments and gather many of their new products from outside the company. The premise for any robust open innovation program is this: Not all great ideas necessarily start within the company. Why not tap into the collective wisdom of consumers and the population at large?

A terrific book that has influenced my open innovation pursuits is *The Game Changer*. Starting out as a test in increasing and sustaining organic revenue and profit growth, A. G. Lafley and his leadership team at Proctor & Gamble (P&G) began to integrate open innovation into everything P&G pursued, while creating new customers and markets. Along with Ram Charan, they show how large companies such as Honeywell, Nokia, LEGO, GE, HP, and DuPont have embraced innovation, including that coming from outside the company. Innovation is not a separate activity, but the job of everyone in a leadership position and the integral driving force for any business that wants to grow organically and succeed on a sustained basis. The world today is experiencing unprecedented change, increasing global competitiveness, and the genuine threat of commoditization. Innovation is the only way to really win market share, and companies do not have a monopoly on creating it. Thank you, A.G., for pioneering open innovation as a corporate discipline.

There are other benefits to a strong open innovation program, not the least of which is that research and development costs can be kept down while the flow of outside ideas inevitably sparks internal innovation as well. The main challenge that most corporate open innovation programs face, however, is they are typically overseen by career employees subject to internal management restrictions, profit motive, and timeline pressures. There is a great deal of financial risk for companies pursuing disruptive, external technologies, and many companies are not willing to take on that risk without close oversight and in-house management. The challenge is that without innovative, out-of-the-box thinking and some reasonable tolerance for risk, many open innovation programs fail. At Lifetime Brands, we wanted to avoid those mistakes. The novel outside-contractor approach we pioneered turned out to be the answer.

Why My Unique Hero-Model Open Innovation Works

The three companies I have helped oversee open innovation programs for have all hired me as an outside contractor. This model has many wonderful benefits. First, I am not an employee and, therefore, am free to operate outside of internal company pressures. Second, I am not restricted by company protocol or guidelines. This is important because it allows me to build strong relationships with inventors, as I am accepted within the innovation community and can directly relate to their challenges. Third, I put a friendly face on companies that innovators have historically distrusted.

Many corporate employees have never owned their own business and are uncertain where to look for outside innovation. I vet things independently and make sure both companies and innovators are protected. What's vital is having a unique skillset developed from an entrepreneurial background that allows me to relate simultaneously with company management and inventors, gaining both sides' trust through sustained credibility. Because this hands-on approach takes a great deal of extra effort, this type of program is referred to as the "hero model" of open innovation. I'm the hero.

10,000 Hours

In his 2008 book *Outliers*, Malcolm Gladwell wrote, "Ten thousand hours is the magic number of greatness."[2] He wrote about Bill Gates, who coded through the night while he was a teen in high school. He also wrote about The Beatles, who played eight-hour gigs every day in Berlin nightclubs long before they invaded America. They became phenomenally successful from spending ten thousand hours developing their craft. Few shortcuts pay off in life, and even fewer in the innovation business. I have put in my ten thousand hours in business building, product vetting, and serving on the front lines of many innovative product launches.

A great book that has influenced my open innovation pursuits is *Outliers*. I would highly recommend reading all of Malcolm Gladwell's books, but if I had to choose one to start with, it would be *Outliers*. The book takes the reader on an intellectual journey through the world of the best and brightest, the most famous and the most successful, spanning a variety of pursuits and industries. The question is repeatedly asked: What makes high-achievers different? One takeaway is that we pay too much attention to what successful people are like, and too little attention to where they are from—their cultures, their families, their generations, and the idiosyncratic experiences of their upbringings. Along the way, Malcolm explains the secrets of software billionaires, what it takes to be a great soccer player, why Asians are good at math, and what made The Beatles the greatest rock band. He also describes the concept of ten thousand hours: Most highly successful people have spent over ten thousand hours honing their trade, so what sometimes appears as luck or overnight success is anything but.

Inside Lifetime Brands

I'm in my fourteenth year of working with publicly traded Lifetime Brands (LCUT on the NASDAQ Stock Exchange). They are the largest non-electrics housewares company in the United States, with 40,000 SKUs and close to 2,000 employees. The company has grown from scratch in the 1960s to almost a billion dollars annually today. Lifetime Brands not only owns the brand name Farberware, they either own or license forty other well-known household brands, such as KitchenAid, Sabatier, and Mikasa. They are the number one seller of kitchen utensils in the world and the largest cutlery company in North America. Chances are you have a Lifetime Brands product in your home right now.

Before coming on board as an outside contractor, it was vital for me to make sure we established certain fundamentals building the program. We agreed upfront to always treat inventors fairly, and that I would be involved hands-on with every outside product submission that comes to the company. I would explore and vet every lead, including both the person and the product. Even if I liked the product, if I found the inventor was close-minded or had irrational expectations, we would move on. Experience has taught me that if folks are rude, or tone-deaf, or not well-prepared, the chances that an eventual licensing agreement will ever be reached are remote, so why bother investing a lot of time and effort. I will not waste the busy people of Lifetime Brands's time on red herrings.

Protecting the company is a big part of my mission. I find lots of products are not ready for prime time or have been done before. Not being prepared hurts an inventor's credibility, so inventors need to take time to prepare before submitting. If I find a worthy product, and the inventor is open to civil discussion and listening, I will present it to the division heads and advocate for both the product and the inventor.

After stringent product reviews, I reconnect with all inventors within a reasonable amount of time to either work toward a licensing agreement or considerately turn them down. I spend a vast amount of time educating inventors on the real path to a licensing agreement and stay involved until we sign and seal the deal. (I will cover more about navigating the submission process in later chapters.)

Lifetime Brands has nine different divisions, and I work directly with the heads of each one. When we put the program into place, I knew we had to build everything from the ground up to create a strong foundation. After extensive research into other open innovation models, I knew several fundamentals had to be respected in our agreement, or I could not participate. My reputation in the inventor community must always come first. I made sure that the inventor's ideas were protected from workarounds, so we agreed the program would not serve simply as an idea-generating design platform. If we did not have an interest in the product, we would not pursue it. For example, if an inventor submitted a revolutionary new garlic press and Lifetime Brands turned it down, then a month or two later introduced something similar, it would destroy any credibility that I built. We also agreed that if an inventor submitted a product idea and Lifetime Brands turned it down, the inventor had the right to take it to another company, and I could continue to help them. This ensured the product would get an in-depth review, and if it were turned down, there had to be a good reason. The Lifetime Brands Open Innovation Program was unique from day one.

We then turned our focus to one of the essential aspects of licensing—royalties. I was adamant from the start we pay the highest royalty possible to any inventor on every licensing deal, provided the cost of goods and the suggested retail was priced right to sell lots of units. Additionally, inventors would not share any of their royalty with me. Plus, inventors would never pay for any service that

Lifetime Brands would provide getting to market, an essential ingredient in making the open innovation program a success. Inventors had to know they were submitting to an honest end-user program and would be treated with respect and fairness. To me, this was the heart and soul of the program.

For full disclosure, everyone at Lifetime Brands trusts my instincts and expertise. I have the green or red light on all submissions as they come in. I get many thousands of submissions every year. If the inventor has not taken the time to get the product ready and conduct their due diligence, I have the 100 percent okay to turn it down on the spot. Unprepared products and presentations will not make it past the gate. Everything from poor sell sheets, not producing a functioning prototype, or the absence of a definitive patent strategy all qualify for not being ready. No matter what the marketing gurus suggest, a drawing alone will not make it through. A product must be well thought out and ready to present. The saving grace is I try to go the extra mile behind the scenes to help inventors get prepared. Even if Lifetime Brands turns the product down, whenever possible I try to lend a hand with products I believe in and steer them in the right direction or provide reasons why the product will not make it.

HONEST TIPS

Ten essentials to the Lifetime Brands Open Innovation Program:

1. Respect every inventor who submits a product for licensing review.
2. Communicate with all submitters in a prompt and courteous manner.
3. Never design around a product submission.

4. If not interested in licensing the product, try to explain why not.
5. If interested in licensing the product, offer the highest royalty possible.
6. If advancing, walk the inventor through the licensing process and contract.
7. Allow inventors leeway if they have built a small business that does not compete.
8. Allow inventors an opportunity to buy direct from LB factories if not disruptive.
9. Give credit to the inventor whenever possible.
10. Stay in contact with the inventor to ensure ongoing cooperation.

Inside Techtronic Industries Power Tool Group

Being a leader in the open innovation arena, I am often asked to speak at conferences around the country. I was invited a few years ago to a large event, which had a fair number of industry experts speaking on outside product development and licensing. The speakers were a collection of brilliant individuals, many of them PhDs. Speaker after speaker presented complex theories of open innovation throughout the day, providing book-smart spiels, though none had ever run an actual, front-line program. I was scheduled to speak in the middle of the day, and when it was my turn, I went up on stage and said, "I feel a little intimidated today because everybody's an academic open innovation expert, but I run a highly successful, hands-on open innovation program." I could immediately see that I had the whole room's attention. There were many company executives at this show, and I was the only one speaking about a real-world program.

At the end of my speech, everyone gathered around, asking questions. I could see two gentlemen standing politely off to the side, waiting to talk. After everyone dispersed, they stepped up. "Hello, we work for a company called TTI."

"Who's that?" I asked.

"TTI stands for Techtronic Industries, the manufacturer of Ryobi, Ridgid, and Milwaukee power tools." They were at the conference representing the $2 billion a year Ryobi and Ridgid Power Tool Group division. "We would love you to develop and oversee an open innovation program similar to the one you've helped start at Lifetime Brands."

This quickly turned into an incredible opportunity. Lifetime Brands's products were geared more toward the female marketplace, while power tools are more of a male-oriented purchase. This would allow me to focus on another exciting industry and the best of both worlds.

"Before you sign me up," I told them, "why don't you host me at your headquarters in Anderson, South Carolina, to spend a day with your management team." I needed to see their operation, get a better feel for their company and fully understand what I could do to help. I ended up spending an entire eight-hour day on-site with their product development team. They listened carefully and asked many questions. Soon enough I agreed to come on board.

Inside Merchant Media

So now, I was involved with the open innovation programs for two large market leaders: housewares and hardware. One industry has predominantly female consumers, and the other has a large percentage of male consumers. I looked at thousands of products in both sectors and helped negotiate licensing deals on behalf of inventors and companies alike. I also helped establish the inventor pavilions at both the national housewares and hardware shows, as well as the PGA Golf Show. (More on those efforts in a later chapter.) I implemented and set up the inventor submission portals for both companies, which were guaranteed to be inventor friendly. I continue to look at every single product submission personally and provide feedback, advice, and guidance.

Recently, I was with my longtime friend Michael Antino, president of Merchant Media, a leading company in the Direct Response Television (DRTV) marketplace. Merchant Media has fifteen different brand names of successful products they've launched, including Smart Spin, Perfect Pasta Maker, Perfect Pancake Maker, and True Touch. These products have been huge successes. He asked me to help find innovative products for Merchant Media as the DRTV space is a vibrant one. I established a program with him featuring the same principles as Lifetime Brands and TTI. All three run slightly different open innovation programs, but all share this guiding mission: Never charge inventors and treat everyone with respect.

I know I am truly fortunate to be where I am today with an incredible opportunity to work with excellent companies bringing exciting new consumer products to market. It has been an honor to work in this arena. I have reviewed well over 100,000 products, participating in hundreds of meetings with company leaders and product development experts. You might say I sit in both teams' huddles—inventors and companies.

2

Who Exactly Are Inventors These Days?

The inventor community is evolving in many positive ways. It's getting younger and more diverse and inclusive. We can now add makers, designers, and product developers to today's ranks of innovators. Historically, we have perceived inventors as hard-nosed intellectuals with unique engineering skills, such as Nikola Tesla, Eli Whitney, or Alexander Graham Bell. We sometimes recognize them as peculiar, introverted, and, in some ways, mad scientists. Traditional inventors seem to have had extraordinary focus and determination, yet many were not attuned to social trends or focused on clear consumer benefits. Many younger inventors today could learn from their predecessors' rigorous standards, but they also have a deeper understanding of the changing marketplace trends. The challenge is to combine the best of both worlds.

Historically, inventors have made a significant impact on our society and are responsible for developing many products and services that we take for granted. Inventors have profoundly influenced the growth of the American economy by solving problems, bringing creativity and competitiveness to the marketplace, providing immeasurable convenience, creating jobs, and enhancing lives. Our founding fathers designed the patent system to be an integral part of our country's DNA. Article 1, Section 8, Clause 8 of the US Constitution protects an individual's right to file for a patent as a property right. It is intended "To promote the progress of science and useful arts, by securing for limited times to authors and inventors the exclusive right to their respective writings and discoveries."[3] To clarify, the clause specifically states that individuals' intellectual property, including patents, copyrights, and trademarks, is a tangible property right, no different than owning land.

The first US Congress wasted no time passing the Patent Act of 1790, placing the responsibility for reviewing each patent application with the secretary of state and acting attorney general, Thomas Jefferson, an inventor himself. As filings dramatically increased, Jefferson recognized he did not have time to consider every patent application. With the aid of Congress, he changed the patent system in 1793, modeling it after England's. Then, any inventor submitting the necessary paperwork, along with $30, plus a declaration showing their invention was new, could apply for a patent. In the beginning, patent applications were not examined by any officials, only registered. To handle the increased demand, the US Patent Office became the second office building built in Washington, DC, after the White House. For over forty years, anyone who applied could acquire a US patent because the process was affordable and quick. Over time, problems developed when inventors swearing oaths that their inventions were brand-new and useful turned out to be complete frauds. After a lot of expensive, time-consuming lawsuits and litigation, Congress placed a middle step between the inventor and awarding a patent, and thus the first patent examiner was hired.

HONEST TIPS

Patent Act of 1790

The Patent Act of 1790 was the first intellectual property statute passed by the US government, about one year after the Constitution was ratified and a new government was organized. The law was concise, defining the subject matter of a US patent as "any useful art, manufacture, engine, machine, or device, or any improvement not before known or used." It granted the inventor "Sole and exclusive right and liberty of making, constructing, using, and vending to others."

President George Washington began signing official patents from the day he took office. He recognized the importance of having a fair and robust patent system that safeguarded the rights of independent inventors. During President Washington's first state of the union address on January 8, 1790, he called upon Congress to develop and establish an official patent system by saying, "The advancement of Agriculture, Commerce, and Manufacture, by all proper means, will not, I trust, need a recommendation." After a few months, Washington signed the Patent Act of 1790 into law. Fast-forwarding two centuries, in June 2018, our country celebrated the ten-millionth utility patent issued by the United States Patent and Trademark Office (USPTO), an event I attended.

The first patent was issued on July 31, 1790, to a gentleman named Samuel Hopkins for a device to make potash and pearl ash. The second patent issued was to a man named Joseph Stacey

Sampson. Mr. Sampson's patent was granted for a method to manufacture a candle. Ironically, a fire destroyed his original patent. The third official patent was issued to a man by the name of Oliver Evans on December 18, 1790. The patent declared a new method to manufacture flour and meal. Mr. Evans's system worked so well that George Washington himself licensed the system and used it with his gristmill to produce spirits and upgraded the entire system at his Mount Vernon home and mill to a fully operational distillery. Which went on to produce over 11,000 gallons of whiskey, making it one of the largest distilleries in America.

Old-School Inventors

The difference between today's inventors and yesterday's is significant. Many inventors of the past seemed to push engineering and technical solutions to the limit. They toiled in labs, basements, and garages building prototypes while carefully maintaining inventor notebooks and ensuring their inventions functioned. They worked hard, sometimes for years, on a project, rarely worrying about marketability. They focused primarily on engineering challenges and performance. Their experience became a guiding influence, as older inventors began integrating their product development skills with a thorough understanding of the patent system. The information they gleaned about engineering and mechanics was utilized to develop and file specific utility patent claims. These detail-oriented values and traits were passed down from our forefathers, ensuring that product development and innovation remained vital throughout our country's history.

Traditional inventors respected inventing and performed a thorough job in every stage of product development. They utilized knowledge and persistence, as many were mechanical-minded engineers who understood the application of mathematics and physics. There was also a great deal of intellectual curiosity as they established new claims to solve everyday problems. One drawback, however, was many older inventors didn't have a refined sense for either the consumer benefit or potential market size of the problems they were inventing a solution for, toiling without an understanding of consumer demand. This led inevitably to plenty of marketplace failure.

I receive many product submissions from old-school inventors, and I respect their process. Some products are brilliantly engineered; however, many are not invented with consumer benefits or sales appeal in mind. Additionally, many traditional inventors are not adept at taking their products to market. Though many have invented great products, they do not know how to license them, or otherwise access retailers and store shelves. Soon, they begin searching for a company, individual, or entity to help them accomplish the marketing task—and therein lies a big problem. Gullible, old-school inventors are easily misled by astute marketers who freely take their money without any reliable chance of marketplace access, let alone eventual profit and sales success.

Today's Inventor

Today's younger inventors bring a revived spirit of innovation and have a better understanding of the consumers they are developing products for. They are also familiar with the five-star product evaluation surveys recently associated with many online retailers and services. They've grown up understanding a well-educated and demanding public. Today's young inventors are also a more diverse group who hail from around the world—more people are entering the innovation space and striving to take inventing to the next level. Instead of doing the same thing over and over, this diverse group of inventors is realizing the freedom involved in being individuals and thinking outside the box. Technology has opened doors and allowed inventors to demonstrate this individuality in unusual ways. They often recognize unexpected problems that require solutions and see holes in the marketplace from a new perspective. In many ways, new inventors represent a breath of fresh air.

On the other hand, many young inventors might fare better learning what traditional inventors bring to the table, including engineering knowledge, prototyping skills, attention to mechanical detail, and keener patent awareness. I find younger inventors adept at utilizing modern software tools such as CAD (computer-assisted drawings) and developing graphic design images and sell sheets. Yet, many omit the technical development of specific patentable claims, which defines the unique function of their new product, enabling them to profit from their ideas. I do find that younger inventors see the world more collectively, which can be a good thing, although they often bypass awareness of their individual rights, which can harm them financially. We'll talk extensively about licensing later in this book, but the reality is many naive young inventors are not as prepared as they should be and, just like their forefathers, are easily misled by clever marketers.

HEROIC INVENTOR

Josh Malone

Josh has terrific engineering skills, works diligently developing products, and has become a leader speaking out for inventor rights. Josh is the inventor of toy sensation Bunch O Balloons. He came up with the idea to help his kids fill up a hundred water balloons in less than a minute. It eventually became the most popular toy in America, selling many millions of units. However, before Bunch O Balloons even came to market, another company copied it and started selling its own versions of the product. Luckily, or so he thought at the time, Josh filed a patent application for his Bunch O Balloons invention. He did not know the enormous amount of time, money, and luck it would take to enforce patents that successfully covered his invention. Eventually Josh settled his patent dispute favorably and the product has gone on to sell millions of units.

Women Make Terrific Inventors

Over the years, I have personally witnessed more and more women emerging as terrific inventors. I believe women have a more intuitive understanding of the consumer they are inventing for and a greater appreciation for the product benefit proposition. It sometimes seems that men attempt to solve technical challenges while women better understand the end use of the product. Women, for example, seem to know that a car is a vehicle that enables them to go places and accomplish tasks. Men more often see a car through its physical characteristics, such as horsepower and wheelbase. Women are far more in touch with the consumer marketplace and pay closer attention to crucial selling details, while men sometimes invent solutions for problems that don't even exist. As women feel more comfortable and accepted pursuing engineering careers, it can only help the world of innovation. Today only 8 percent of all patents are issued to women as the sole inventor—this is a statistic that needs to change dramatically.

Children, Teenagers, and College Inventors

What I love most about the innovation world is that doors can open to anyone with a great product. It doesn't matter if you are a male, female, minority, young, old, your national origin, or your stature. I have been impressed by product submissions from many young inventors across the country, including those ages six to eighteen. It is encouraging to see the American spirit of innovation being fulfilled through these young minds. However, many submissions from this age group are not fully developed or thought through, sometimes because of limited life experience. Specific industries may be more intuitive for them, such as toys, sporting goods, apps, and software. Young inventors are quick to leverage new tools and technologies and pay attention to the unseen opportunity. Being nimble and open-minded, though, doesn't always outweigh experience. I encourage young inventors to keep in mind that licensing is a business, and even young inventors must prove concept when they pursue patents and licensing.

Makers

Makers are different from inventors in many ways, although they are related. This group is a little more like the people I knew back in high school; we called them "shop guys." Remember back when men were encouraged to take shop class and women home economics? Men were expected to work with their hands. They built furniture and engines, always studying the inner workings and developing new ways to repair them or make them work better.

Makers are the hands-on folks who are willing to wade into uncharted waters and explore things from the ground up. The primary difference today is this community is more tied to advanced technology such as CAD and 3-D printers. Makers have embraced innovation and are making products with modern tools. In retrospect, we have shifted from *Popular Mechanics* magazine to *Make* magazine. There are now Maker Spaces all over the country, allowing affordable access to specialized equipment. The American spirit of ingenuity and innovation thrives in the minds of these quietly brilliant people, while they explore and produce new products organically. The only thing I worry about is that many makers don't understand the patent system and are not aware of how to properly monetize their creations. I feel both groups, makers and inventors, can learn a great deal from each other.

HONEST TIPS

Makers and Maker Spaces

The Maker Movement is a social phenomenon with an artisanal spirit, emphasizing learning through doing. Dale Dougherty, the CEO of Maker Media and publisher of *Make* magazine, is often credited with being the father of the Maker Movement. Maker culture encourages novel applications of technology, as well as the exploration of ways to work such as metalworking, calligraphy, filmmaking, and computer programming. The rise of the maker culture and maker fairs is closely associated with the rise of maker spaces throughout the country and world, which are community locations that charge monthly fees for training classes and use of high-tech equipment, such as 3-D printers.

Product Developers and Designers

Product developers and designers are also an essential part of today's innovation community. Some work alone, while others are part of a larger company or team. Sometimes they generate their ideas, other times they are hired by individuals or manufacturers to take a concept from general description to specific ideation. I receive many submissions from these talented people. Designs alone, without a technical component, are hard to advance for licensing. Utility patents typically have more value than design patents. Also, contrary to what some middle-man invention marketers advocate, simply combining multiple existing functions into one product garners little interest from companies—at least the ones that I work with. Remember, larger companies often have their *own* in-house design teams and are already working on blue-sky development projects, so they are looking outside the company for more disruptive solutions, ones they have not previously considered.

3

Bridging the Cultural Divide Between Corporations and Inventors

All companies are different. This includes the industries they are in and the makeup of their senior management, board of directors and shareholders; not to mention, their sales volume, number of employees, financial resources, market share, brands, distribution channels, and product assortments. What they should all have in common, however, is they are made up of a collective group of diverse personnel who work together to generate profit.

Some companies operate in traditional ways, utilizing standard methods taught at business schools, while others take more novel approaches that reflect the self-interest of their management and shareholders. Companies, culturally, are group enterprises, not to be confused with the independent, self-interest culture of inventors. Collectively, companies perceive things differently than individual inventors. So, if you are an independent inventor trying to work with a corporation, you need to identify what core beliefs govern and motivate the company you are approaching. Whether you are submitting your product for review, negotiating terms, or signing a mutually profitable licensing agreement, you need to understand where you and your product fit within the company mission.

Some Companies Embrace Open Innovation

Some progressive companies recognize this reality: Not every good product in the world originates from inside their four walls. Though many companies have full-time product development teams of intelligent employees working on potentially prolific projects, those employees are often focused on their own daily routines and company protocol. They generally respond to retail buyer interests instead of organically solving consumer problems. Most companies don't have the time or manpower necessary for deep-dive inventing, which can be a distraction. However, there is a whole world of original ideas, concepts, and product development strategies provided by independent innovators that can be available to companies if they are willing to open up to the outside world. The real question is whether the company has the will, means, organization, and confidence to pursue them.

HEROIC INVENTOR

Thaddeus Alemaeo

Thaddeus, a mortgage broker by day and longtime inventor on the side, developed the Odor Absorbing Splatter Screen™. The screen is a terrific kitchen utensil, which not only protects the stove and cabinets from hot frying pan grease, it eliminates odors for those who may not have an overhead exhaust vent. Lifetime Brands licensed this patented product and has gone onto sell millions of units over the years. It's been a terrific example of how companies and inventors alike can win when working together.

Other Companies Are More Closed-Minded

Other companies are reluctant to open their doors to outside innovation. Why? There are a whole host of reasons that I have personally encountered. One, they often don't understand or trust people outside their four walls. Many company employees have grown up within the organization and know little beyond what they've learned there. Most employees want to do their job, collect a bonus or raise, and possibly be promoted; however, they often lack the entrepreneurial background to liaison with lone wolf innovators. Therefore, inventors need to be aware, when presenting their product, of the need to overcome such natural bias until credibility is established. Inventors need to prove they are professional by describing clearly how their submission benefits the company and translate how licensing their product will connect in the marketplace with consumers and make the company more profitable.

From time to time, some mid-level managers have issues with inventors potentially earning more money from royalties than they do. This bias is subliminally embedded in different ways, including the perception that the inventor is lucky, rather than good. What the mid-level managers don't understand is that the inventor has probably pursued dozens of other product failures before becoming an overnight sensation. It requires senior management to keep things on track in the licensing relationship, someone with vision who is looking out for the greater good of the company.

A significant pushback to any outside submission platform might also arise from within the company's legal department. Nothing personal, but corporate lawyers are as concerned about the adverse effects of inventors suing or harassing the company as naive inventors are that companies will unfairly steal their ideas. Call it mutual fear. It's often easier for the legal folks to shut down all outside interaction regardless of the effect it may have on a company's sales or growth unless they are specifically instructed to open up. Because of this, some companies publicly announce the fact they do not accept outside submissions. When you observe this, it's typically based upon a legal perspective. Attorneys have been assigned the task of protecting the company, even if it means putting an artificial curfew in place. There is often an immense cultural divide between those outside the company walls and those inside. Often, they do not have the collective experience or game plan in place to allow outside contact, even if it means passing on potentially terrific innovation. It's simply safer to build a wall and shut the gate.

Opening Up to Inventors Requires Corporate Commitment

Some companies see the open innovation process as a huge hassle because it means putting corporate lead-collection and management systems in place, dedication to marketing, regular product reviews, and outside communication. Open innovation also includes product development, research, field testing, adding features, and predicting the future, which involves both anticipating and embracing change. I like to call this the innovation tumult. Without people getting excited about and embracing new and disruptive ideas, it's tough to advance outside products internally. Where there is risk involved, it usually takes a confident champion to stick their neck out and lead.

For companies not collectively bitten by the innovation bug, this can be challenging.

Any good company that wants to pursue outside innovation needs to embrace change, which takes time, commitment, and resources. Yet, many people within companies do not really want to deal with change, so there may be internal politics to sort through. Unless company employees taking risks are presented with a vision of potential reward, a half-sustained open innovation effort will soon falter. Open innovation is a mindset that needs to be implemented and reinforced from top management down. Without company-wide support, efforts will fall short. Everyone needs to believe in it.

Within large companies, employees are faced with many distractions. Their daily work involves meetings, deadlines, demanding bosses, sick days, and politics. Most strive to perform well; however, it is hard enough to deal with everyday issues, making it even more challenging to focus on new projects that may or may not provide profit down the road. It's like drafting a teenage baseball player for the big leagues. There isn't any guarantee he is ever going to make it, so why put energy into development with only so much time in the day? Unless the excitement of a new product exceeds the daily workload, or management puts a premium on innovation, things are not going to advance. Internal self-interest and survival often trump risk and potential. What a company needs to see is a winner, and then everyone will better understand the program's benefit and purpose.

There are also, of course, those unscrupulous companies who have no interest in rewarding anyone outside their walls for innovation. They believe they can do it better themselves, or, if something superior is presented, they can simply take the idea or design around it. Unfortunately, some companies think such an approach reduces their risk. However, this approach defines them as rip-off artists, and word travels fast about such companies. Maybe the company is an original pioneer in a certain category, or management believes everything should come through them. The philosophy is to let others innovate and enter the space later to reap the spoils when a product becomes successful. Why pay the upfront cost of innovation when it can be copied or stolen? Unfortunately, this happens a fair amount in the direct response TV industry. It's challenging to identify these companies, but seeing how a company treats inventors is a good place to start. Ask around within the inventor community, at local inventor clubs or trade shows, about predator companies. You'll find out quickly who they are. Be sure to avoid them.

Big versus Small Companies

Implementing an open innovation program takes an immense amount of work, team building, vision, and coordinated strategy. Smaller companies may find it especially challenging to put an open innovation program in place. Large companies have a better chance of building a program because they have the ability to afford it. Most large companies are also under additional pressure from shareholders to sustain growth. They typically need to introduce new products to generate additional profit. When a company dedicates a department or staff to accepting outside product submissions, it is a good sign for inventors.

A company's ability to invest in a product roll-out is also essential, favoring those who have more financial and product development resources. If you are offered a licensing agreement by a smaller company or individual, research their new product launch history, capacity, and credibility. Look at their planned path to market, including existing retail relationships and brands. Do your homework. If everything goes well, can they truly afford to invest enough for you to be highly successful?

Sometimes Companies are Working on the Same Product as You

Sometimes companies are already working internally on the same problem that an inventor is, so it is not uncommon to receive competing product submissions in the same space. The winner will be the inventor with the best technical solution at the most affordable price. If the outside product is superior, it's a quick fix for the company, and all is good. Yet, if the company pursues a different avenue, inventors must understand it's their right to pursue that course, assuming they arrived at the other solution completely separately and ethically.

The Cost of Mistakes

The cost of making mistakes is the real trade-off for companies when deciding upon establishing an outside innovation program. Can the company suffer failure and still move on to develop the next product? Do they have the resources, stamina, and desire to handle short-term challenges to find the big winner down the road? These decisions are usually made at the top of the company. An inventor needs to identify visionary corporate leaders who choose to pursue outside innovation without fear. A company's open innovation approach is reflected in its reputation and stable of innovative products. When thinking of large innovative companies, I think of 3M, Procter & Gamble, and Lifetime Brands. All three companies have legendary leaders and management teams dedicated to open innovation success.

HONEST TIPS

Ten reasons why the outside contractor model of open innovation works well:

- Prevents the distraction of internal personnel from outside badgering.
- Puts a friendly face on the company to foster early outside trust.
- Gains far more leads than those that naturally come into the company.
- Fully engages the outside innovation world with appropriate protection.
- Utilizes the resources and knowledge of an outside guide, not someone company trained.
- Utilizes an expert in advancing negotiations and explaining licensing details.
- Uses a dedicated person to advance NDAs, agreement protocols, and contracts.
- Inspires a greater spirit of innovation within the company as outside leads flourish.
- Uses success stories to build a stronger program and more future leads.
- Uses the outside contractor to pursue other projects of special interest quietly.

The Independent Contractor Model

I have served as an outside contractor overseeing several open innovation programs, bridging the gap between the inventors and companies. There is a great benefit to this approach for most companies. My role protects all parties and allows me to vet outside products before presenting to the companies I help. It allows me to serve in an unbiased role that respects both cultures. The companies I work with trust my judgment, and inventors also trust in my ability to understand them and speak for their interests—a win-win.

Along with trust comes the avoidance of legal issues. In Malcolm Gladwell's book *Outliers,* research reveals that 90 percent of all medical malpractice suits would never have been initiated if the doctor hadn't been a jerk. The same is true of running an open innovation program with an outside contractor. A nice face for the company, including plenty of personal empathy and trust, goes a long way toward establishing the company's reputation for innovation. This is important when working with creative people.

It has been surprisingly easy for me to get along with the employees of all the companies I have helped over the years as an outside contractor. It's simple: I take a big distraction off their shoulders, and they usually don't see me enough in person to really annoy them. Internally, companies are sometimes reluctant to absorb the extra work involved in reviewing external leads because it takes manpower to perform the required diligence on every single lead, and there can be thousands of them. It is better to have an outside contractor dedicated to such filtering. Submissions will come into companies anyway, no matter what happens—even if the company puts a big sign that says, "No Outside Submissions." I simply address them. I also communicate regularly with inventors, which is typically music to their ears. I repeatedly hear how companies ignore them, that their calls and emails are never returned. This is not good. It builds distrust for companies, which can result in many unintended consequences.

All of which leads me to this theorem: When bridging any two cultures, with sufficient foresight, human consideration, trust, and appropriate infrastructure, the positive results from effectively combining two worlds usually outweigh the limitations of one.

4

Developing Your Product (Part 1)

Taking the Right Approach as an Inventor

Aspiring to become a great inventor takes persistence, professionalism, and the right attitude to succeed. If you want to have a better chance to achieve integrity and success, you will greatly increase your chances if you take a serious, detailed, and comprehensive approach.

Professional inventors seek solutions to problems or try to improve the quality of life. To become profitable, however, the invention must solve a problem big enough for enough consumers to ensure it becomes a commercial success. To get you started, begin asking yourself this question: Does my invention truly solve a big enough problem for enough people? This will help you decide on whether to pursue your idea, or which path you take to market, either licensing or going direct. By the way, allow me to add here, if your plan is to improve upon a current product, the added features must be significant, or the price point far lower, otherwise you will be perceived as an also-ran.

If you choose the licensing route, many units must be sold, assuming a modest royalty amount is paid on each one, to earn real money. What's real money? For a $20 or $30 product, a nice financial win is when at least tens of thousands of units are sold. But for a real hit, we are talking about hundreds of thousands or even millions to generate significant profit. These types of hits don't come along every day. Of course, the product must also be manufactured at the right cost to sustain margins. And it needs mass appeal, which is supplemented with promotion, word of mouth, and positive online 5-star reviews. This is something larger companies can help accomplish.

The other scenario to achieving profitability is selling fewer units of a niche product but at a much higher retail price point. With this scenario, licensing may not be the best path to market. The better route might be going to the market on one's own, making a higher profit margin per unit on fewer sales. Of course, there can be more risk to this approach. We'll discuss both paths to market, licensing and going direct, in upcoming chapters, but regardless of your choice, your product, and a rational plan to exploit it, needs to be developed.

Research the Marketplace

Once a new product idea has gotten you excited, the first step is to thoroughly vet the marketplace in your targeted field to see where your product fits in. This will provide critical feedback and should come before spending a lot of money finalizing prototypes—and certainly well ahead of filing for a patent. I am a firm believer in learning about comparable products and retail price points, as well as specific consumer trends and preferences.

Start your search by identifying related products and who manufactures them. Where do you go to conduct your consumer research? The easiest way is to start online. Begin with simple keyword searches for similar products. Go on Amazon and other retailers. I am always amazed by the number of submissions that I receive where a similar product is already on the market, which tells me the inventor has not done their homework and is probably wasting my time. These online retailers are a global marketplace right at your fingertips. You can begin a simple search by typing appropriate keywords in their search box to see what products pop up. It's never been easier. You might be surprised at the extensiveness of results. Then expand your search with added features and details as you scroll through each product and learn more. Studying customer reviews lets you know what real consumers think about the quality and usefulness of the products. The reality is that if a product gets less than four stars from the Amazon consumer reviews, it will not last long on the market shelf. Negative reviews, through word of mouth, will quickly kill sales. Honest customer reviews are probably the most significant advancement in the past ten years to the consumer shopping experience. As a product developer, take advantage of this easy way to gather insight, because if similar products do not work well, it may be your cue to pursue a better solution.

Next, visit retail stores and speak with salespeople. Ask questions of those who are knowledgeable about the category. Then, subscribe to trade industry magazines and, if you are able, visit an industry trade show to see firsthand what similar products are already on the market—plus discover any new ones about to be introduced, which can be an incredibly valuable experience.

INVENTOR INDUSTRY ROCKSTAR

Toy Industry Designer Matt Nuccio

Matt Nuccio has a wealth of knowledge in the fields of design, engineering, prototyping, licensing, and sourcing; he is passionate about helping companies and inventors convey a clear message through design. His design and development agency, Design Edge Inc., is a multi-award-winning international development agency with offices in New York and Hong Kong. For over twenty-five years, Matt has been designing products and packaging for hundreds of companies big and small across many industries, including toys, electronics, food and beverage, health and beauty, houseware, hardware, and pets. He has also written many columns focusing on design, inventing, engineering, and manufacturing for *Toy & Family Entertainment*, *Toy Book*, *Royalty $*, and *Licensing Book*. Matt is also a frequent lecturer on design and manufacturing, having spoken at trade shows, inventor clubs, and universities around the world. A recognized subject matter expert, Matt was elected co-chairman of the Toy Industry Association associate panel representing all designers and inventors in the toy industry. He also generously gives of his time through the United Inventors Association, helping especially with social media efforts and educational webinars.

Patents and the United States Patent and Trademark Office

Ground zero of the patent and intellectual property arena in the United States is the United States Patent and Trademark Office (USPTO), located in Alexandria, Virginia. They have an in-depth website (www.uspto.gov) providing inventors around the world with many methods to research and save any publicly disclosed, filed patents.

Searching for and extensively reviewing existing patents to determine if there are any previously claimed physical and design features that you are working on is an especially important next step. This "Prior Art Search" must be done with diligence, patience, and persistence. You can begin by searching the US Patent and Trademark Office website or hire professionals who specialize in such searches. Patent attorneys also do this—but understand that, behind the scenes, they typically hire independent intellectual property search specialists themselves, some of whom are based near the USPTO in Alexandria to research patent validity for them. These searches form the basis for opinion letters they are asked to write sometimes when a patent application is filed for and potential companies or licensees want more confidence there are no obvious infringement issues they might have to deal with. These letters are not cheap.

Establishing Your Intellectual Property

When licensing, the importance of having IP protection is essential. It forms the tangible basis for your legal agreement. If you file for a utility patent, you need to decide if it's a twelve-month provisional patent application (PPA) or a full non-provisional patent application, which will be formally reviewed on the path to approved patent status. Note, a PPA will need to be converted into a full non-provisional filing before signing a licensing agreement with most companies. If you haven't established company interest within a year, you will need to reevaluate whether your product has any market value and, if not, the PPA filing is automatically dropped. When a product comes across my desk that includes patent filing information, it moves up the ladder of importance. Why? Because I know that I am dealing with an inventor who has put effort into developing the product, and it is likely more than just another idea.

Intellectual Property Protection

Patents are an integral part of the product licensing process, representing the defined property rights licensing agreements are secured to. Even though our patent system may be under assault today from larger tech-firms and USPTO PTAB rulings, intellectual property protection is still essential.

Within the intellectual property arena, there are seven primary types of patents issued by The United States Patent and Trademark Office:

- Utility Patents
- Design Patents
- Provisional Patents (Subset)
- Non-Provisional Patents (Subset)
- Plant Patents
- Business Method Patents
- Software Patents

Provisional and non-provisional patents are filed for and issued in the process of pursuing utility patents. Utility patents are the most valuable type of intellectual property when licensing your product technology. There are also other types of IP, including copyrights, trademarks, works of art, and famous sound recordings. There is even intellectual property protection, in various degrees, for secret ingredients or formulas such as Coca Cola, Kentucky Fried Chicken, and Pepsi.

Utility Patents

When an invention features functional components for a new and useful purpose, manufacturing process, the composition of matter, or even new and useful improvements, it is of value to obtain a utility patent. The utility patent is the most widely used and valuable patent from a licensing standpoint. Once issued, it lasts up to twenty years from the filing date of the application. Two other patents, *business method patents* and *software patents*, are sub-categories of utility patents and have come under fire in recent years. Business method patents, for reference, are patents directed toward new and specific methods in conducting business. These types of patents are difficult to obtain because it is often particularly challenging to precisely distinguish a new way to conduct business. Software patents are another sub-category of the utility patent issued for new and improved software and its interaction with computer hardware.

Utility Patent Claims

Patent claims are what come from building a functioning prototype, the critical process that allows you to observe all the physical mechanisms necessary to make a product work. The information gleaned from your observations will establish the claims of your invention, which are the essential portion of any utility patent and make your invention unique. I am not sure how anyone can file a legitimate utility patent application without understanding the inner workings of their physical prototype. The best way to determine specific patent claims is through the trial and error process of building a prototype, which is why this process is essential. Many utility patent filings feature multiple claims; in fact, they may have dozens.

Design Patents

Design patents are for the "ornamental design" of an invention. Design patents provide patent protection on specific designs of an invention and last for fifteen years from the date the patent is granted. Design patents are useful; however, they are harder to enforce, from an infringement standpoint, than utility patents. Therefore, they usually command lower royalties when licensed.

Provisional Patent Applications (PPA)

Provisional patents are an easy aid to assist independent inventors. Here are a few insights into the history of the provisional patent. The provisional application was introduced to US patent law in 1994 by way of an amendment to the Patent Act of 1952. The United States provided twelve months to foreign-filed applications (this had been in effect in the United States since 1901). During this time, there was a US ratification of the Brussels revision of the Paris Convention for the Protection of Industrial Property. The provisional patent was the answer and the domestic filing equivalent to the twelve-month benefit that had long been given to foreign applications.

One year should be ample time to vet your invention and potential licensees. However, the twelve months should also be used to narrow any specific mechanical patent claims about your invention. Where the provisional patent has inherent value is placing you farther ahead in line for a strong utility patent application when you are ready to convert to a non-provisional filing. It plants the "first to file" flag. In March 2013, the Leahy-Smith America Invents Act (AIA) was introduced, which prompted the United States to switch from a "first to invent" to the new "first to file" rule, stating that whoever is the first to file a patent on an invention has priority in owning the rights.

Currently, in the United States and the world, whoever files first has the rights to that invention. The PPA allows you to be the first to file for far less money and effort than a full non-provisional filing.

A provisional patent application should be written to include as much information on your invention as you can provide. Every nuance and feature should be described in detail. Do not leave anything out—the smallest of details could be the difference in your licensing success. Some companies and coaches encourage the practice of putting a quick PPA together to get it filed. Once again, this is usually the wrong advice. Number one, a PPA should be written as if it were an official utility patent application and its filing taken seriously. Remember, a strong, well-written PPA could be the foothold to a formal utility patent application that could be filed once the twelve-month timeframe has lapsed. Number two, if you enter into a potential negotiation of a licensing agreement, you want to be confident that everything is in place with your invention. A PPA should stand on its merit. For example, imagine sitting in a boardroom with executives from a large company, and you are meeting to discuss a rather sizable licensing agreement for your invention. Everything is going as planned until the dreaded question is brought up, "Do you have a patent?" Your answer is, "Yes, I do. I have a provisional patent application on file." A lump now begins to grow in your throat because you remember the half-baked, poorly written, provisional patent application you sent in. "May we see your provisional for review?" is the next question they ask. Your spirit sinks because now you must submit an inadequately written PPA that will not hold up. And once credibility is tarnished, things tend to head south.

A PPA is a statement of the invention with a declaration of claims and a clear description of what it does, showing how it solves problems better than any other inventions like it. A PPA is expected to contain information similar to a non-provisional utility patent application. All PPAs should start with an official written application, which is an application for a patent that shall be made or authorized to be made by the inventor or inventors. The written portion will specify the ongoing title of the invention, describing in detail exactly what the product does. According to USPTO Title 37—Code of Federal Regulations Patents, Trademarks, and Copyrights, a provisional application must include a cover sheet identifying the following:

- The application as a provisional application for patent
- The name(s) of all inventors
- Inventor residence(s)
- Title of the invention
- Name and registration number of attorney or agent and docket number (if applicable)
- Correspondence address
- Any US government agency that has a property interest in the application[4]

Next on the list is your *contents*. The contents section is the summation of the specifics of the invention, a drawing or drawings if you have them, and an oath declaring your invention is new and improved. Provide every nuance and detail, leaving nothing to doubt or misinterpretation. This is especially important. The next essential is *claims*. These are the most crucial part of your application that defines the scope and extent of your invention. Claims contain everything significant about your invention, in technical terms, how each feature functions as well as the overall operation of the invention. This is the portion of your invention considered the intellectual property and what you are seeking to protect from infringement or liability.

The next section of the PPA is the *filing date*. This is straightforward and is the date of your application, with or without claims, received by the United States Patent and Trademark Office. Since the system is now "first to file" instead of "first to invent" the date is critical as it serves as your "stake in the ground" toward your claims. The last step is the required *authorization*. Every provisional application is made to be authorized by the inventor granting access to file.

A PPA is inexpensive to process, in the range of $400 to $1500 depending upon legal help, but only good for twelve months from the date of filing. PPAs are never published and not officially reviewed by a patent examiner. They serve as a placeholder serving the first inventor to file requirements. I highly recommend using a qualified patent attorney to draft and file your non-provisional patent conversion. The stronger the patent, the more valuable it is, so make sure not to cut corners when you convert after twelve months.

Worldwide Patent Filings

There is another patent route an inventor can take, which is becoming more popular: filing a PCT, a worldwide patent placeholder. It is also called a Patent Cooperation Treaty patent application. The Patent Cooperation Treaty is an international treaty that declares patent rights granted between contracting nations, which include most of the developed countries in the world. A PCT is a single application filed in any international receiving office. The United States Patent and Trademark Office is an official receiving office here in the United States. A PCT gives the inventor or company that files the patent application priority in the future with any other contracting country. It is commonly referred to as an international patent application. Once an inventor files a PCT, now that America is part of the Worldwide Patent System, the United States will honor the patent filing. This enables an inventor to return and file the non-provisional version within eighteen months. This was a side benefit of the Leahy-Smith America Invents Act. PCTs may be helpful for inventors and companies on their way, for whatever reason, to worldwide patent status. But if your plan is to pursue business only in the United States, they are a luxury.

When a patent, whether a non-provisional or provisional, is filed, the filing allows you to work under secrecy and have protection. This is vital to the process because your invention will not be revealed to anyone that you don't allow to view it. Remember, once eighteen months have elapsed, the patent does become public and can be read by anyone. A non-provisional patent application, unless the applicant files what is called a non-publication request, as with all US Patent applications, will be published automatically after eighteen months from the earliest priority filing date, even if it is not yet granted.

Searching Google Patents and Images

Google has developed a user-friendly patent searching system called Google Patents (patents.google.com). This system includes every patent ever registered with the USPTO. Use the same keyword techniques with Google's patent search system as you would with the USPTO system. There is the slim possibility you could even uncover a patent not found within your USPTO search. Record your findings of similar patents and evaluate your idea against others in the marketplace.

One method I recommend is using Google Images. Keywords trigger the imaging system on any specific subject. When reviewing a new product, type in the same keywords used in an Amazon search and, within seconds, images of similar products are listed. The photos are all linked to specific webpages, where products may be further investigated. Incorporate this step into your market research. It might surprise you what it reveals, which could be critical to your decision on moving forward. Make note that the images of the products you find may or may not be on the market. I recommend clicking the pictures of specific interest and examining the product in other retail-friendly places to see if it's already on the market.

Patent Drawings

Your research should also include studying patent drawings. This will provide insight into how inventions work mechanically, presenting a clear perspective of all the functioning parts while helping you to define the claims of your individual patent. Understanding the mechanics provided in the drawings will also help you design and make your prototype. As you study the patent drawings, make a note of any elements that could apply to your invention. There are professionals who specialize in these types of drawings.

Importance of a Functioning Prototype

As mentioned throughout this book, a functioning prototype is a critical component in the product development and licensing submission process. Even if you do not need it on day one, it will be necessary if there is company interest in your product. When it's called for, if you don't have one, it's been my repeated experience that the submission process starts spiraling downward. A working prototype reveals a lot about your seriousness as a product developer: Please don't be misled by marketers saying all you need to license an idea is to mail in a sell sheet with a picture or sketch. This will make your odds for successfully securing a licensing agreement much longer. Even if you are lucky enough to advance without a functioning prototype it will reduce your royalty rate. A prototype is a real, working sample, not a drawing or CAD animation. It can take time to build a prototype, but it allows for "proof of concept," shows functionality, and demonstrates how the problem is technically solved. It doesn't have to look great; it just needs to work.

Please get in your garage, lab, basement, or workshop—it doesn't matter how or where—and discover what's working and not working. Find practical answers to your challenges. If you don't know how to build a prototype, research the field, and find someone who can. This is the same premise as hiring a patent attorney—or a plumber, for that matter. If you can't do something well, don't let it stop you, find someone who can help. I'm not advocating haphazardly spending money. If you genuinely want access to the licensing game, think through what it means to become financially successful. Talk to other seasoned product developers at your local inventor's clubs and associations, which are located all over the country. Search for local prototype companies or connect with universities or colleges in your area. Please don't waste your money on anyone who advises that an attractive sell sheet is good enough. Even if you are lucky enough to get in the door with a sell sheet, the next question will surely be, "Can we see your prototype?"

So, now that you know, be ready.

A great book that has influenced my open innovation pursuits is *Bad Blood* by John Carreyrou. It is the story of Silicon Valley-based biotech startup Theranos, founded by Stanford dropout wonder child Elizabeth Holmes. She invented a small portable device to conduct a multitude of blood tests without hurting the patient with needles when drawing blood. A great idea, but Theranos never built a unit that functioned properly! Hundreds of millions of dollars down the tubes later (and court cases still to come), it has become one of the epic inventor disasters of all time. Every inventor should read the book to understand better why building a functioning prototype matters.

Retail Price Points

During your research, make a note of the retail price points of similar products. This information may be helpful when negotiating a licensing agreement. If the retail price point of a product is too high, the product will not sell, which means companies will not pursue a licensing agreement. I often ask inventors, "At what retail price point do you think your product will sell?" Many don't have a clue. Having the answer, by the way, affirms that you have performed diligence and pushes you further up the ladder in my estimation. Estimating potential manufacturing costs is a good place to start. The suggested retail price is, in most instances, four to five times the initial factory cost, depending upon the industry. Or, vet the retail marketplace for items of similar dimension, materials, and function to see what they sell for and divide by five.

Materials

Keep in mind throughout your market research what product materials are necessary, which will help determine the functionality of your prototype—and, ultimately, the cost to manufacture. Learning about different materials, and when they can be substituted for, may bring your costs down and improve performance. Such research can also provide insight into fabricating your prototype when proving function. Materials might be plastic, metal, wood, carbon fiber, rubber, or foam, as well as many others.

Product Sell Sheet

A product sell sheet should be an essential part of your product submission efforts. It should never be longer than one page, or you will quickly lose the reviewer's interest. (Remember, the reviewer is probably busy with many other submissions.) The sell sheet should be simple, meaning as few words as possible, and include a photo and a description of the significant benefits of the product. Make sure the product sell sheet is attractive, well designed and worded, and conveys critical product selling and benefits information. A link to a video or website is always helpful as well, as is information about intellectual property status and your contact information.

A Video Tells a Thousand Pictures

There is another advantage to building a working prototype; you can record the product in action. I can't overemphasize how helpful videos are these days when reviewing a new product. You might have heard the expression, "A picture tells a thousand words?" Well, I like to add, "A video tells a thousand pictures!" Some companies I have worked with will not even review a product unless there is a video. I often see confusion when an inventor attempts to verbally describe their invention. The invaluable understanding that comes when a product is seen in action is lost without video evidence. Animated demonstrations simply do not carry the same weight. They usually prompt the next two questions: Is that real? Have you built one? A video not only shows your product's functions, but it also creates excitement. All videos should be short and concise. It doesn't have to be an elaborate, expensive production. You can even make a simple video on your phone, just keep it under a minute. Why? Most people who review products do not have the time to sit through a lengthy five-minute video. So, quickly show the benefits of your product and then trust that an experienced person on the other end will get it.

My Advice on Improving Your Odds for Success

The difference between a novice and a great inventor is taking the right steps and going the extra mile, which will improve your odds for both licensing and marketplace success. Many inventors take shortcuts. They seem to get tired, or are lazy, or are distracted working on too many other mediocre projects. Or maybe they are coached poorly and just start mailing it in. It's quite easy to tell on my review end when this happens. Product development is more than designing a sell sheet and contacting as many companies as you can find in the phone book. Professional product development and inventing involve thorough marketplace research, building a functional prototype, continual refinement, clever problem solving, defining substantial claims, and filing strong patents. Make sure you understand everything about your product and the industry you are attempting to exploit, and make certain your product solves a real problem that appeals to enough paying consumers. Remember, the licensing profit game is not finished when an agreement is signed; it is only the beginning. Licensing revenues derive from sustained sales success, so any shortcuts in development will not, in the long run, make it past observant consumers. Sorry to say, but real product development that has a fighting chance of advancing into a functionally disruptive product at market involves hard work.

When an inventor submits a product to me, I know very quickly whether they are serious or get-rich hopefuls. In fact, I can usually tell by the time I finish reading their online submission form, which is when I ask for information about where they are in the product development process, including prototypes, patents, costing, and marketplace vetting. If there is interest in pursuing a product, I prefer talking on a professional level with the inventor directly about their product. When I do, I don't like to hear the words, "I think it should work." Or, "The company can take the development process from here." This half-baked approach does not cut it in the real world, inside successful companies. If you are not prepared, your product will either be shot down or—even if you are very lucky—your royalty rate will be significantly reduced. Whenever those happy press releases run about a licensing agreement being signed, have you noticed we never see the royalty rates or other specific terms? We also rarely see a follow-up press release a year or two later describing how the product is selling. I wonder why? Could it be a poor agreement was signed, or no real sales success was achieved, or the product was never manufactured because it did not work?

Be advised. There is a lot of marketing misinformation that floats around our industry, primarily for the benefit of those interested in collecting money from you and making it all sound easy. Stay focused, guard your money (as I like to say, "If you don't want to get ripped off, put Grandma on the cash register"), and invest wisely in developing your product the right way.

5

Developing Your Product (Part 2)

A More Structured Approach to Ideation: The Scamper and TRIZ Methods

There are many ways to generate new product ideas. Some happen organically, while others are developed through strategic formulas, methods, and calculations. The organic approach to innovation may be satisfactory for most independent inventors. Companies, however, often take a different approach to generating marketable products. The methods described in this chapter represent a more structured approach, which many inventors may find very revealing and could learn something new from. Try for a moment to look at product development from a professional perspective. For corporations, the term "innovation" should be described as, "The *profitable* implementation of a new idea," as innovating is not only about problem-solving but also about making money. This is not as easy as it may sound.

Professional Idea Generation Methods

Independent inventors view the world from their own perspective and, to their credit, often observe problems others don't notice. It's typically a more organic approach. Companies, on the other hand, with multiple employees and lots of infrastructure expenses, pursue product development more by team and formula.

I would like to provide an inside view of corporate product development labs, where two of the most prominent ideation methods are:

- SCAMPER
- TRIZ

As an independent inventor, you should be aware of these two systematic approaches and personally consider learning more about product development from them.

The SCAMPER Product Development Method

SCAMPER is an advanced idea-generation technique used by individuals and corporations to generate marketable ideas. SCAMPER is an acronym for a checklist designed to be a creative thinking tool. SCAMPER helps individuals to either create changes to an existing product or to create new products from scratch. Bob Eberle designed SCAMPER initially to allow children to tap into their inherent creativity. As a result, the acronym is simple to aid in the memorization of the steps. Many companies are now using the method for strategic enhancement of product platforms as well as disruptive product development.

- The SCAMPER approach is more of a methodology used to create ideas and alternatives until you arrive at a functional product. SCAMPER begins with *substitute*, which means to interchange components or materials to make a product better. Can the product be used elsewhere or substituted for another idea? Are there other ways you can use it in the product? What would happen if you changed the purpose or function of the product?
- Next is to *combine*, which refers to mixing, combining, and integrating parts of a product with other assemblies or services. Ask what could happen by combining this one idea with another. Or what significant benefit would be created by combining two different products. Ask what you can combine or bring together somehow to make the product more functional. Can you combine purposes or ideas?
- Following is *adapt*, which asks if the product can be altered, functionally changed, or used with any other part of the product to adapt. To adapt, ask what can be shifted to provide a better solution? Is there anything else like this? What could you duplicate or emulate to adapt? What can be changed on this product without losing its original function?
- After that, there is to *modify* by increasing, changing, or reducing the size of the product in scale. This is important with a lot of variable questions to ask. Here you could ask, can you change or modify the product by changing the color, size, shape, smell, or form? Modify also leads to questioning, can you magnify or minify? Can you make the product larger or smaller, lower or higher, and maybe even lighter or shorter?

- Following modifying is to *put to another use*. This refers to if the product can be used in other ways than the intended function. What other markets might this product sell in? What would work best if it were used in another setting?
- Then there is *eliminate*, which is often overlooked yet impactful. Ask what you can take away to make the product better. What is it that can be reduced, shortened, or minimized to save waste, time, or costs? What can be simplified, removed, or reduced down to the core to improve the product?
- Lastly, there is *rearrange*. Is there anything that can be interchanged, such as parts, components, patterns, or even layout? What could possibly be turned upside down or switch positions, shapes, and directions?

To use SCAMPER successfully, you need to take the time, slow down, and ask these basic questions to generate creative thoughts for more predictable results. Today, many large companies, universities, organizations, and inventor groups use the SCAMPER method to create insightful and marketable product ideas. As an inventor, try using the SCAMPER method when you ideate and see if it helps.

INVENTOR INDUSTRY ROCKSTAR

Product Developer Trevor Lambert

Trevor is president and founder of Lambert Licensing, one of the leading invention agencies focused on licensing and technology transfer. An inventor himself, after seeing deficiencies in the business model of existing invention service providers, Trevor established the company with the goal of providing inventors with improved transparency by working on a commission basis. Trevor then founded an additional design and engineering firm, Enhance Product Development, Inc., that specializes in providing inventors with tailored services for both licensing or manufacturing and distribution. Enhance provides turnkey services for inventors from industrial design to prototyping to domestic or overseas manufacturing sourcing. There are even design packages specifically for inventors seeking to license their inventions, particularly in the DRTV space. Trevor has been a longtime supporter of inventors, donating both his time and resources to the United Inventors Association.

The TRIZ Method of Product Development

The TRIZ method is an even more sophisticated and powerful tool in creative idea generation and problem-solving. TRIZ was developed in the U.S.S.R by Russian scientist Genrich Altshuller. Many large corporations use this method to develop strong, repeatedly marketable ideas that help solve identifiable problems. This method targets forty distinct principles that contribute to strategic product development.

For starters, the organic brainstorming of ideas is only effective if you have thorough knowledge about your category of interest. The challenge with brainstorming, however, is that it's often unreliable and unpredictable. This is where TRIZ excels. The TRIZ method is based upon existing data, research, and logic rather than merely on gut instinct and brainstorming. For companies and individuals, the use of TRIZ makes problem-solving and product-generation substantially more predictable and reliable.

In layman's terms, the process builds upon problems already solved in the past. The theory is, whatever problem you are experiencing, someone somewhere has already tried to solve it . The key is to locate a solution and apply what was learned to your current development process. Here are the critical TRIZ tools that will help you in understanding this method of product development:

- Generalizing problems and solutions
- Eliminating problems or "contradications"

The inner working of TRIZ begins with the theory of problems and solutions having been repeated (or invented) across multiple industries, practices, and sciences:

1. Problems have a way of repeating themselves across different industries and sciences.
2. Patterns have repeated across various industries and sciences.
3. Problems are "contradictions" and solutions can be predicted for that problem.
4. Innovations use scientific methods outside of specific fields they were developed for.

To use TRIZ in your product generation endeavors, you need to learn and recognize the repetition of the patterns of problems and solutions. TRIZ involves understanding the problems within any idea-generating scenario and using improved methods to discover different varieties of the problem, providing a definitive solution. The essential aspects of TRIZ are as follows:

1. Define your problem
2. Locate the TRIZ general problem close to or matching it

3. Find the generalized solution solving the problem
4. Adapt the solution to your specific problem

Most ideas are born from problems of past ideas. The problems are either physical or technical. With TRIZ, take away the flaws, and you can solve the problem and make a marketable and reliable product.

INVENTOR INDUSTRY ROCKSTAR

Bob Hausslein

Born and raised in New York City, Bob, in his youth, attended the famous Bronx High School of Science. He then went on to study for many years at MIT, eventually receiving a PhD in chemical engineering. Bob worked at Polaroid, helped launch two startups, and has filed twenty-five US patents, plus some foreign equivalents. I met Bob through the Inventor's Association of New England (IANE), where he ran one of the best inventor clubs in America for well over twenty years. A wonderful thing was Bob's close association with MIT, which allowed the club to meet monthly on campus. So, I was always able to say in jest, after speaking there many times, that I had lectured at MIT. Bob is one of the most sincere and generous human beings that I have ever had the good fortune of meeting, either inside or outside of our industry. He tirelessly donated his personal time and resources, without ever charging anyone a dime, educating many inventors in the Boston and greater New England area. To me, giving back as Bob has for such an extended period of time is the essence of American goodwill and philanthropy. IANE is now led by another terrific person, Bob's hand-chosen successor, George Peters, so the club will thankfully remain active and helpful to yet another generation of aspiring inventors.

TRIZ Forty Principles Outlined

TRIZ is a Russian acronym for the Theory of Inventive Problem-Solving. The method is based upon five significant product-generation principles and forty innovation principles used to address problems and solutions by lateral thinking. TRIZ presses any inventor to view problems and products differently from a diverse perspective. These solutions were discovered by studying thousands of patents as companies use these principles to solve problems and develop products across many industries. The key to TRIZ is that it is universal and was designed to provide more predictability in innovating products. Large global companies such as IBM, Boeing, Google, Dell, and many others use TRIZ today to create new products. Yet, the same process can be used by you, an independent inventor. The use of TRIZ will help you learn how to optimize the product development process but also help you gain a deeper understanding of product trends in the marketplace. TRIZ, and the many tools and procedures it offers, has quickly become the go-to choice for the development of innovative product concepts.

You can use the principles in any combination to help ideate and create. Remember, as an inventor, following the same systematic process of product development as large corporations will only push you further by developing products with predictable results.

TRIZ includes five key principles essential to understand. It all starts with looking at *the end result* by first thinking outside the box. An inventor would do well by changing their mindset and not being satisfied with the first solution to a problem. Allow yourself to be open to better ideas and other possibilities. Another principle is *less is more.* As independent inventors, you need not invest sizeable sums of money to achieve the best product idea. You can invent using materials you already have and get the same or better solution without spending thousands of dollars. The key to TRIZ is principle three, which is that most, if not all, *solutions already exist.* Using TRIZ will help you define problems by examining general principles that allow you as an inventor to find solutions outside of your product's category or field. Another essential principle of TRIZ is *searching for contradictions*, which supports the mindset that inventing is problem-solving, which exists through contradictions. By defining the contradictions, we usually discover a solution. The most important principle of all, which makes TRIZ so effective, is called *the lines of evolution.* With TRIZ, the understanding is that systems and processes do not happen randomly. Methods are produced by repeated fixed patterns, which makes product development entirely predictable.

The forty innovation principles of TRIZ begin with *segmentation*, which describes how to divide an object into individual parts. By doing this, product pieces can be examined and assembled or disassembled, as necessary to create the solution. Next is *extraction*, which is used to remove or retrieve any part from an object. You would do this by studying the object and removing only the necessary section. The third section of TRIZ is *local quality*. This portion is used to examine whether the product parts are the same, not heterogeneous, and here local quality is used to differentiate whether the parts carry out multiple functions. For this to be effective, each part of the product must be placed under conditions that allow it to operate at full capacity. *Asymmetry* is the next step, where the product is looked at for its asymmetrical properties, which would help the product function. *Consolidation* is the next step, where a product and its parts are consolidated to work more efficiently in its environment, space, and time. *Universality* is determining whether a product can perform multiple different functions and removing elements that have a negative outcome on the operation of the product. Ingenuity is involved with the next step, which is *nesting*. This is where one object is placed inside another product to make it function. The key to nesting is these steps are repeated if necessary, to make a fully functioning product. *Counterweight* is the process where the weight of an object is used combined with another object to compensate with another object that creates a lifting force. *Prior counteraction* is the process of loading counter tension to an object to test for stress points that could cause the product to not function or fail. *Prior action* is any necessary changes made to a product before its operation. *Cushion in advance* is where you reinforce the reliability of the function of the product by taking precautions in case of an emergency when the product is in operation.

Next is *equipotentiality*, which is merely changing the functionality of the product so that it will not require lifting or lowering during operation. *Do it in reverse* is where a product's function is studied by implementing the opposite problem on it. For example, if a product heats, then reverse the task by making it cool instead. Or turn the product upside down if it is meant to be right side up and see how it functions. *Curvature* is the step to evaluate any surfaces of the product. If the surface is flat and the product does not perform adequately, use curved parts such as rollers, balls, or spirals and replace linear movement with centrifugal force.

Dynamics are vital to the function by looking at the product characteristics to see if anything needs to be changed for optimal performance at every stage. Look at every part of the product and, if necessary, change the position for better functionality.

Partial or excessive action is where a product has difficulty reaching the maximum performance, accepting the level of functionality. Evaluate if the product is acceptable at the less desired result you're looking for. *Another dimension* is challenging but necessary to evaluate. A product needs to transition from one-dimensional to alternate dimensions such as two or three dimensional. Next in TRIZ is *mechanical vibration.* This is referring to products such as MRIs, X-rays, and other mechanical products. So, if your product uses oscillation, evaluate if the oscillation needs to increase to ultrasonic. The product's mechanical vibrations could be replaced with piezo vibrations or ultrasonic waves.

Periodic action is the next step where you replace a continuous movement or action of a product with a regular motion by changing the frequency. If this is the case with your product, you could create pauses between impulses to provide additional action.

We are halfway through the TRIZ process, and number twenty is *continuity of useful action.* This is the step where all parts of the product should continuously operate at full capacity. Examine if all parts function and carry out their function without any failures or breakdowns. *Skipping* is where the product's function malfunctions at remarkably high speeds. To coincide with these factors of TRIZ, next is *blessing in disguise: "turn lemons into lemonade."* Take any negative portion of your product and increase the function to such an extent that it ceases being harmful or dangerous. *Feedback* is next and is a simple process. Introduce feedback through using output instead of input to improve output checks. *Intermediary* is next, and this is where you can merge one product with another product temporarily.

Number twenty-five is *self-service*, which means making a product serve itself by offering essential self-performing functions. A self-checkout machine you find at any supermarket is a prime example of self-service in action. *Copying* is using an inexpensive replica instead of a valuable or fragile product. Using cubic zirconia instead of real diamonds is a good example here. In the same mindset with this principle, *cheap objects* is the use of disposable or inexpensive materials to reduce costs. An example here could be plasticware or plastic cups. For *replace the mechanical system* use alternative mechanisms that function by altering or changing mechanical products. An excellent example of this would be folding chairs or ladders. *Pneumatics and hydraulics* switch out solid parts of a product and replace it with hydraulic and pneumatic devices. A doctor's table or dentist's chair that rotates and elevates through hydraulics are good examples.

Next is an odd TRIZ principle called *flexible shells*, where a product can deliver results by being changed by an outside coating. Products such as sunscreens, breathable shirts for golf, and sports slacks are prime examples of flexible shells. *Porous materials* are products that use specific elements that are not easily damaged. You can find these porous materials in everyday products such as bandages and gauze used in healthcare. *Color changes* are the method used to make products that alter their colors, such as polarized sunglasses and automobile glass and windshields. *Homogeneity* is the process of making different products interact with one another. Industrial paints and adhesives are great examples of this in action. *Discarding and recovering* is the process of taking products at the end of their useful life and reusing them for other purposes.

Parameter changes are products that function by changing the essential properties of the product. A steam shower is a perfect example. And if you have ever seen a home pregnancy test, this is an excellent sample of *phase transitions*. These are products that experience and offer a solution through chemical changes. The sun and solar panels are growing extremely popular and when it comes to TRIZ, *thermal expansion* is a prime example of a product that converts heat to mechanical energy (electric).

Dr. Scholl's product for wart removal is a sample of *strong oxidants*. This process involves reinforcing the oxidative process to produce a specified function. Next to last, we have *inert atmosphere*, which consists of switching a neutral environment to enable a direct and wanted feature of a product. And anyone who has been camping or hiking has probably used products that involve the mat principle of *composite materials*. This is the process of combining materials to make a product function. Have you ever drunk from a thermos or a whiskey flask? If so, you have experienced the benefits of using composite materials.

6

Going to Market (Part 1)

Licensing Your Product

Now that you have developed your product, how will you profitably take it to market? Choosing the right path is an important decision. Will you license through others, or go directly on your own? In the next few chapters, I will discuss the pros and cons of each approach, as I have had significant experience with both. Let's begin with licensing, which for many reasons, is the path most inventors choose.

What Is Licensing?

Licensing is working through others, known as licensees, including companies, individuals, and independent entities, to take your product to market. Royalties are paid from the licensee to you, the licensor, on units sold, typically including a minimum number of unit sales per year and possibly an advance payment. A licensing agreement can be exclusive or nonexclusive for the licensee; however, exclusive deals command a higher royalty rate. Licensing contracts might also include specified geographic territories (such as North America or worldwide) and specific trade categories, such as commercial or consumer. Licensing agreements are generally legally tied to intellectual property, including patents and trademarks.

With most licensing agreements, the percentage rate of royalties paid is based upon the strength of the intellectual property and the cost to produce the product. If the royalty amount adds too much to the base cost of the product, the retail price may need to increase, which will slow down sales significantly, in which case no one benefits. There are different ways to profit from licensing agreements, but the best way is for the licensee to sell a large number of units of your product.

The Advantages of Licensing

The primary advantage of licensing is that inventors have an opportunity to earn money with less personal financial risk. When companies license a product, they invest their own money into creating, marketing, and selling a substantial number of units. Not only do they serve as the manufacturer, but they also play the role of an angel investor. The amount of money a company invests in developing a product and taking it to market can be significant.

There are other advantages to licensing. In addition to marketing your product under their brands, industry-leading companies may have allotted shelf space within large retail stores gained through years of strong relationships, outstanding products, and even slotting fees.

Along with shelf space and financial support, a company can also help make your product better. Many companies know far more about their industry and what consumers and retailers are demanding in real-time, plus they have the ability to lab- or field-test products to improve technical aspects that you may have been unable to observe or finalize. Larger companies usually have the resources, both financial and manufacturing-wise, to tool-up for production and work through early design challenges. Companies will also develop attractive, educational packaging, instructions, and point of sale materials that help explain the product at retail, while managing the fulfillment of large quantities of goods through warehouses and onto retail selling floors. The process of getting products from factories to the retail shelf is complicated and expensive. Better companies have well-established distribution channels.

HEROIC INVENTORS

Adrienne McNicholas and Michelle Ivankovic

Adrienne and Michelle are the creators of the wonderful kitchen product, Food Huggers, which are patented, reusable silicone covers that preserve the freshness of leftover fruits and vegetables. From the beginning, their goal was to design the best possible product and improve sustainability by reducing food waste at home, while cutting back on landfill overuse and wasted money. I first met them via Kickstarter, where they launched a highly successful crowdfunding campaign. What proved exciting was tracking them down in Europe and establishing a licensing agreement with Lifetime Brands. Adrienne lives in Madrid and Michelle in Amsterdam, yet they were able to collaborate on the design and engineering from afar, not to mention work with me. Food Huggers is now a product platform at Lifetime Brands, with over twenty SKUs sold under multiple brands in different countries.

Identifying Inventor-Friendly Companies

Many inventors distrust large companies for fear those companies might steal their product and leave them out in the cold. Most of the time, this is an irrational perception based upon hearsay and lore. However, it is not entirely without merit. The critical thing for inventors licensing their product is to determine the credibility of any company they wish to approach. Eventually, it's unavoidable to share information. This can be a scary step. From the inventor's perspective, most companies are made up of people they have never met before, with unknown motivations. How you identify and deal with this uneasiness may determine your chances for licensing success, so identifying friendly companies is important.

Conducting due diligence and vetting companies that are a fit for your product begins with the submission process. The first step is determining whether any company you aspire to approach is even in your field. You need to research the assortment of products they produce. It is surprising how many inventors do not do a thorough job of this, as well as the number of submissions I receive for products that are not connected to the companies I work with. This will usually result in a quick rejection, which leads some inventors to ask what the company is looking for. At this point, I know they are not ready for prime time because they have not done their vetting homework. Once I get this sense, I wonder if they are applying the same laziness to their entire product development process.

Researching Reputable Companies

The easiest way to research a company is to go online, do a keyword search, and read carefully through company websites that come up, as well as any relevant articles. Sometimes financial journals can also provide helpful insight into a company. Walking retail stores and inspecting the packaging of similar products will supply a great deal of company information, which you can use back in your office to pursue further. If it's possible, attending a trade show and visiting a company booth is the best way to cover a lot of ground in a short period of time, but it is not always convenient. You can, however, sign up for trade industry magazines to find specific company listings. It's actually easy to search and discover companies in a new arena. You do not need to pay a marketer to acquire an impersonal telephone directory of names. I know from first-hand experience that contact information offered in some of these generic listings can be dated or inaccurate. Do your own research. The key is being intellectually curious and expending some good old-fashioned investigative work.

Online Researching Google and Amazon

A Google search can provide a quick company snapshot with pertinent information such as annual sales, target markets, CEO, and influential people within the company. If you are searching for a company without a significant market presence, you may not find them—and yet, in its own way, this may give you insight into the reach of that company. Within Google, continue your search with broader keywords and expand in concentric circles.

I receive thousands of product submissions every year and quickly eliminate 50 percent of the submissions because the inventors haven't performed any research on their ideas. How do I know this? The first destination I often visit is Amazon. I type in general keywords to see what products come up. Amazon is a global marketplace. If a product already exists, it will probably be listed there.

When I review products that already exist, I quickly reject the submission and make note that the inventor does not understand what they are doing, which does not bode well for the next product they come back with. Credibility is essential in the innovation business, and once credibility is lost, it is hard to regain. A simple sixty-second search would have avoided this problem. I recognize some folks may be adamant about their product idea, and I certainly don't want to dampen anyone's enthusiasm, but please do your homework. It will be readily apparent if you don't.

Important product information can also be gained from Amazon by reading customer reviews, which often reveal problems and concerns the consumer public has with a product. This information can help you in the prototype-development stage. A quick search can provide valuable insight into product functionality, benefits, pricing, and materials. If you find a product at all like yours, this is the time for serious evaluation. In some instances—and I know this is tough—finding a similar product that you cannot significantly improve upon may be a heads-up to throw in the towel and move on to another idea.

Attending Industry Trade Shows

I will cover trade shows later in more depth, but here's a glimpse for the serious product developer. As you research potential companies, be sure to get out from behind your desk and attend shows and events that specifically relate to your market interest. Trade shows are where almost every company in an industry gathers.

They provide an excellent opportunity to meet and evaluate different company personnel. By attending, you can also take advantage of a wealth of industry speakers, educational forums, inventor pavilions, and trade publications. Be sure to walk into booths, introduce yourself, and ask who handles inventor submissions. Find out if that company operates an open innovation program. Don't share anything significant, such as a prototype, unless you have an NDA signed or have already filed for at least a non-provisional patent. Be diligent, but generally describe what you are up to, discover if there's potential initial interest, and determine whom to contact after the show for a more serious review.

Trade shows are a terrific place to locate inventor-friendly companies and senior executives. When approaching good companies at trade shows, how they respond when inquiring about their open innovation program is meaningful. Ask which of the products they are displaying have been licensed from other inventors and which ones have sold successfully. If they are showing none, that may be a heads up they are not as engaged with inventors as they might have you believe. This information can help you to evaluate the company better. In certain situations, you may ask for the names of the inventors who have had past success, and then speak with those inventors about their experience with the company.

Walking Retail Stores

Visiting retail stores that offer similar products to the one you are working on can be quite valuable when researching which companies to approach. Speak with salespeople to get a general idea if the category you are pursuing is popular, or if it's simply niched. I have found salespeople to be deeply knowledgeable about up-to-the-minute customer trends. Plus, most have no interest in stealing your idea (or ability to do so if they did). Salespeople make terrific sounding boards, particularly if you do not know other professionals in the field. Carefully examine the packaging of similar products and locate the company's contact and website information. You can learn who the major players are by their product assortment on the shelf. Once you have compiled a list of companies from store visits, continue your field research with calls and emails.

Identifying Companies Unfriendly to Inventors

One critical aspect of licensing is determining the right companies to license your product. I will warn you, not every company is inventor friendly. Many companies are simply fearful of working with outside inventors and product developers. They have little interest in signing non-disclosure agreements and want to keep all their innovation development in-house. It does not necessarily mean these companies are troublesome, or that they steal and copy other products in the marketplace. It more likely means their legal people are calling the final shots. Your job is to find out which companies are not willing to work with outside innovations.

There are many companies that do not review outside products at all. They do not want to take on the hassle of reviewing submissions, nor the risk of getting sued for infringement. Because of this, many companies have found it easier to simply say no to all outside innovation. You can find out their company position through research, as many of these companies publish this language right on their websites. Other companies do not accept outside ideas simply because of their company size, resources, and financial status. Many are not able to dedicate the necessary finances to pursue such an endeavor.

Operating an open innovation program takes resources, dedication, and company stability.

Your job as an inventor is to perform thorough research and find trustworthy, reputable companies within your product's category. Here are some early indicators to help determine if companies are legitimate and fair:

- Non-disclosure agreements
- Corporate legal department
- Royalty caps
- Complicated submission process
- Unfriendly staff

Non-Disclosure Agreements

One of the first things to examine is a company's non-disclosure agreement. Just like a stale piece of bread can ruin your meal at a restaurant long before the main course is delivered, a poor NDA can destroy a relationship early on with a company. I have witnessed one-sided NDAs over the years, those written way too much in the company's favor. Note that most companies want you to sign their NDA. There is nothing suspicious with this practice; they typically get so many requests to sign NDAs they can't always take the time to evaluate hundreds of different templates. It's far more efficient to advance one standard NDA. The companies I work for have taken the time to draft fair NDAs for both parties. My own NDAs mirror the NDAs of companies that I help. Since I am an outside contractor, I also need to sign them. Most inventor-friendly companies offer an equitable NDA, which is the first step in determining if they are behaving honestly. Read them carefully or have your attorney review and advise.

Corporate Legal Department

Legal departments within companies protect the company's interests in all agreement matters. In my opinion, they should not be running the open innovation effort by themselves. Though they make certain all legal matters are compliant, they should not make the decision about which products will be successful and be licensed. That's a division-level review. I have found when contacting a company, if they immediately instruct you to submit through the legal department, it's a sign the company is not inventor friendly and has other priorities. Attorneys are vital to the licensing process, but not as the opening move.

Royalty Caps

I have seen inventors sign licensing deals with royalty caps. A royalty cap is when royalties are no longer paid if sales hit a certain agreed-upon amount. This approach is usually the result of a middle-management mastermind who thinks the company can get away with it. If this is put on the table, it's a big red flag, and I would recommend looking for other companies. I have always found senior management much more excited about paying royalties, because it means the product is selling and adding profit to the company's bottom line, regardless of how much money the inventor is making.

Complicated Submission Process

If you find a company in sync with your product, but you discover it is virtually impossible to submit to, this should raise a red flag. Companies that are open to outside products usually offer a welcoming way to accept a product for review. If a company makes it hard to present a product, or the process is tedious and cumbersome, be wary. Use your best judgment on this. It should be evident early on when a company embraces inventors and open innovation. Be sure to submit the right way, follow their protocol, and pitch your product professionally.

Unfriendly Staff

Is the receptionist knowledgeable? It sounds simple enough, but another indication that a company is inventor friendly is when you call and ask about outside product submissions, the phone receptionist knows exactly where, or to whom, to direct you. Conversely, if you call a company and they have no idea where to direct you, it's never a good sign.

Before Submitting Your Product

Before submitting products to a company, inventors need to read the fine print that companies use to describe their open innovation programs. Do your research and learn everything you can about the company. The more you know, the better you will also be perceived by them. Sometimes inventors ask me what type of products my companies are looking for. This question alone raises a red flag because the inventor hasn't done much research. My advice is to go online, visit trade shows, check out catalogs, and visit stores to get yourself up-to-speed on any company's product assortment. Once that's done, if you have a technical question about what's selling in a specific category, you will appear far more prepared and worthy of further interest.

Remember, companies considered "inventor-friendly" usually let it be known. This can be accomplished by placing a positive announcement on their homepage. Some companies understand there isn't a monopoly on good products and want to engage the outside world. With the companies I work with, we steer external product developers to a short questionnaire. Because of the volume of submissions, I developed this system to evaluate each submission more efficiently. I can glean a great deal of information about you and the product in a short period of time.

I share a sample submission questionnaire on p. 79, but here's a summary of information to share. From my inside perspective, the best approach is to prepare well before submitting your product to a company. Even if there is not an opportunity to fill out a company questionnaire during the product submission process, it's still essential to concisely communicate your product when the time is right.

Present your product as efficiently as possible, briefly explaining the primary benefits, features, intellectual property status, where you stand in the product development process, and what personal goals you are trying to accomplish. Describe the problem you are trying to solve and explain the primary reason why you are attempting to solve it, along with your market research and what makes your product unique. Explain your existing utility patent searches and provisional patent application filing status. Confirm if you have engineered plans, any CAD drawings, or professional renderings of your product. Disclose if you have built a fully functioning prototype and provide insight if you have ever manufactured your product, then clarify if you have filed with the USPTO for a specific patent or trademark.

Approach a product submission as if you are applying to college. Admissions counselors learn a great deal about a college applicant from the questionnaire, and it does not take long for an industry expert to understand you and your product and determine how serious you are. I learn more from the answers to these questions than you can imagine. Early in the submission review process is a critical time for me. It's important to make a great first impression, because if I can't understand your product, status, and goals, I may shelve the submission. You don't have to finalize the sale in the first minute, but you should have a goal of opening the door in the future to a more thorough review.

HONEST TIPS

Product Submission Questionnaire

Here's what I ask inventors early in my product review process. These questions have served me well for many years.

- Please identify the problem you are trying to solve.
- Please explain the primary reason why you are attempting to solve this problem.
- Describe any market research you have pursued to date.
- Please list the main feature that makes your product unique.
- Have you conducted a utility patent search?
- Have you formally filed for a non-provisional utility patent? If so, please provide the filing date.
- Do you have engineering plans, CAD drawings, or professional renderings?
- Have you built a prototype? Is it handmade, or professionally constructed?
- Have you ever manufactured a quantity of your product?
- Have you filed with the USPTO for a specific trademark?
- Have you submitted your product for review to other companies? If so, when?
- Is your primary goal licensing, or proceeding to market on your own?

Submitting to Multiple Companies

Contacting multiple companies at once is acceptable when submitting your product for review. There isn't anything wrong, early on, with generating a little competition. If a company gets back to you and asks for some time to evaluate your product exclusively, you can certainly oblige. Because then you can make an informed decision on how best to proceed. Companies should be confident in striking a good deal or otherwise be supportive if you go somewhere else. Companies know one product should not make or break them, and if they need your product, they will know to treat you fairly. In dealing with a company, if you feel pressured or intimidated for any reason, they may not be the right company. If you plan on submitting to multiple companies, look for signs of being open innovation–friendly, such as:

- Do they have an open innovation program and an easy submission process?
- Do they have a dedicated open innovation person or department?
- Is their non-disclosure agreement fair?
- Do they respond on a timely basis?
- Do they have a good reputation within the inventor community?

Sometimes, the largest company in a business category may not be inventor friendly, and if so, cross them off your list. For example, when you visit the Lifetime Brands website, there is a direct link to submit your product ideas. For the inventor, when a company makes its submission portal open and inviting, this means the company is prepared and expecting you to submit your products. However, researching companies doesn't stop here. Look carefully at the fine print, terms, conditions, and NDAs. You need to determine what is reasonable and make the process work for you.

Researching a Company's History of Success

Leading companies should be able to show evidence of licensing success. Besides learning if they run an official open innovation program, find out what products the company has previously licensed. This is a question that I would always be willing to answer. It's far easier giving examples of five to ten licensing successes than describing a dozen divisions, scores of categories, and thousands of products. This isn't information typically available to the public, so inquiring is acceptable. Examine the company's inventor success record, including how many successes there have been to date. A successful company should have a substantial track record. With Lifetime Brands, I have initiated over a hundred licensing deals by treating inventors fairly and offering superior support.

Approaching Large or Small Companies?

One question I often hear, is it better to approach a larger or smaller company? I would always recommend starting with the larger, more prominent company that has established brands, strong retail relationships, industry clout, and a successful track record of launching new products. Bottom line, the larger companies can sell more products, and if you are receiving a royalty for each unit sold, this will help you earn more money. Larger companies are established in more retail outlets and have better brand awareness among consumers. Most large companies have more clout in keeping manufacturing costs low, sourcing goods at the right cost, which allows a lower price on retail store shelves. Most of all, large companies are likely to invest heavily in fine-tuning the product and ramping up inventory production. So, when I hear the announcements from middle-men marketers about inventors receiving a licensing deal, my professional thoughts immediately progress to the size of the company, if the product will be reordered, along with how successful sales will be two or three years down the road, and the company's ability to expand sales.

I suppose it's possible that some smaller companies may not require a functioning prototype. They may be new to the open innovation world, or maybe the company does not have an established licensing protocol. Remember, any company can sign an exclusive licensing agreement and pay a small advance, then shelve your product for up to two years if development and functionality do not work out. Carefully vet every company you approach. Ask what their track record of success is? Has the company signed a licensing agreement before? Have they had marketplace success? Do they have the necessary resources to be successful? Make sure you do your homework.

Approaching a smaller company is not necessarily a poor option, and sometimes it may be your only option. Perhaps a smaller company offers something profoundly unique, or the owner is special, or the product assortment more closely fits your product. Remember, always keep your options open with industry-leading companies until they say no or stop getting back to you on a timely basis. The challenge with small companies is working with quirky individuals who may place their own interests ahead of yours. Or, maybe the company lacks the financial and employee resources to launch products in a large scale coordinated way. Perhaps they oversell their abilities to entice you to license with them. Do as much diligence on every company, large or small.

Look for well-known household, volume-oriented brands that are afforded shelf space by major retailers. Research significant players who sell the most units of similar products. Most of the significant open innovation players will allocate a portion of their website for product submission reviews. There are multiple open innovation-friendly identifiers that you can use to distinguish a good fit for you and your product. When researching companies, look for these characteristics:

- Is the company an industry leader with significant support resources?
- Do they have a dedicated open innovation submission process?
- Do they respond to you promptly?
- Do they have a fair NDA and open to rational discussion?
- Do they offer fair licensing terms?
- Do they have a history of licensing success?
- Overall, do you consider them inventor friendly?

Corporate Brands and Fulfillment

Leading companies typically have a stable of influential brand names that loyal customers recognize and are often pre-sold on. These companies, through purchasing power, can also hit more aggressive retail price points, which ensures the product will sell-through, which, in turn, helps the company negotiate lower production costs. Top companies can seamlessly oversee the entire supply chain cycle, including order taking, shipping, billing, accounts receivable, and reordering, not to mention the handling of damages, repairs, refunds, and returns. As the licensor, you can take advantage of the market power these companies possess. Major companies may also extend your technology into a platform of multiple products as well as market your product under numerous brand names they either own or lease.

Companies Can Help Protect Your Intellectual Property

Another major benefit to licensing is that larger companies with in-house legal departments can assist in patrolling and defending your intellectual property, including patents and trademarks. They can help protect your product from infringers by leveraging their legal staff, long-term industry relationships, and sales clout. When a big company partners with you, they want to succeed as much as you do and will back you up. Plus, other big companies usually don't mess with a marketplace leader. They have too much to lose if the company gets upset and comes back after them. Typically, larger companies understand the turf and generally respect each other. When negotiating a licensing agreement, you should examine the company to ensure they can offer these helpful resources.

One thing I often reiterate to inventors is keeping your licensing options open for as long as possible until you must make a final decision. Remember, no deal is complete until you sign the licensing agreement. Until that time, you need to ask thoughtful questions, rather than demanding unrealistic terms. This measured approach will help you move the process ahead and leave the door open in the future for more product reviews.

One other upside to licensing is most good companies with strong open innovation programs embrace the innovation and creativity that inventors bring to the table and are usually excited to review new products. I represent companies that offer systematically fair submission processes that uphold high ethical standards. Most reputable companies have protocols, rules, and procedures in place to protect everyone. How a company treats inventors says a great deal about them and should answer the question if they are a company worth working with. Do your homework to learn who the good ones are.

Listing Potential Companies in Order of Preference

There are companies of various sizes around the United States and world. On your company list, assess each one in order of size and fit. Bigger companies almost always generate more sales, but at the same time, are they right for you? Leading companies stand out by offering recognizable and identifiable consumer brands, so don't forget you want to work with a company that can get your product on store shelves. Overall, examine companies in your arena and place them in order of whom you feel is the best fit for you and your product. Next, approach each of them in that order, assuming they are inventor friendly. If not friendly, skip them and go on to the next one on your list.

Clear and Concise Licensing Presentation

Lastly, having a concise, readable, product presentation is an essential starting point in the licensing review and submission process. I always appreciate a clear description, even if it's verbal, because I need to quickly understand both you and your product. If I don't fully understand, but still like the product, I will reach out for more information. Conversely, if I am on the fence about your product, I may not. Most products in the middle fade with more examination, so it all depends upon timing. Don't let that happen to your submission.

Wasting Time and Effort

Unfortunately, I receive submissions for products that aren't within my contracted company's arena or expertise. This wastes both my time and yours. Likewise, it always impresses me to receive a submission that refers to a company's product assortment or brands. This shows me the inventor has gone to the trouble of learning about the company. Taking a dressed-up product photo with the company colors and logo added can add a memorable touch. When I receive a product such as this, I will give that inventor extra time and attention. Remember, your intellectual property rights are the physical portion that most licensing agreements are tied into when signing a licensing agreement. Strong patent claims derived from prototypes are essential to improving your royalty rate. Traditional old-school inventors understand the value between both intellectual property and prototypes. Somewhere along the way, the concept of not needing a prototype or a filed patent has crept into the inventor licensing arena. I believe it stems from poor advice, the easy lure of doing things cheaply, and the promoted idea that exerting a little bit of effort and then sitting back and passively earning a living is actually real. Well, if it is at all, it's exceedingly rare. What I can confirm is that any inventor who has built a prototype, proved function, and filed for a utility patent, improves their ranking and credibility, and paves the way for more serious licensing discussions.

7

Inside the Corporate Product Review Process

I have personally helped, in one hands-on way or another, tens of thousands of independent inventors during the past twenty years. I love working with grassroots innovators who are smart, creative, technically intuitive, and essential to fostering disruptive innovation in America. Though most companies may recognize the potential benefit of working with inventors, they also see collaborating with them as a significant challenge. Why? Besides being creative and tech-savvy, many inventors also have quirky personalities and unorthodox business practices. The challenge is bridging the cultural divide while establishing trust and credibility.

Open Innovation as a Corporate Discipline

Companies that have open innovation departments have established them as a business discipline much the same as finance, marketing, or sales. Some companies have immense open innovation divisions with hundreds of employees such as Unilever. Within such companies, the tone and tenor of their open innovation programs are set by senior management, core guidelines are established, and everyone in the company follows them.

Open Innovation Guidelines

When I began discussing an open innovation program in 2008 at Lifetime Brands, there were specific guidelines that needed to be followed to make the process fair:

- How would the company handle product submissions and reviews?
- What ethical codes of conduct needed to be established?
- What kinds of privacy issues needed to be maintained?
- What legal parameters needed to be in place?
- How would the company handle intellectual property protection?
- What types of license contracts would be established?

A company must develop ground rules to ensure everything is fair when interacting with the outside world. When establishing an ethical product review platform, there isn't any room for a lone-wolf employee within the company playing by separate rules. There is a specific submission protocol that must be defined, and the review process must be integrated. I don't find it at all amusing when an inventor attempts to go around me and submit a product to someone else within the company. Everyone at the companies I help understands my role that I examine and vet *all* products before presentation to the division heads. In the end, every product comes through me, which is the protocol agreed upon. Open innovation programs must be run with discipline. When submitting a product to a company, ask if they adhere to such high standards.

Companies' Perception of Inventors

As an inventor, have you ever given thought to how companies perceive you as an outsider? Do they respect inventors, or do they feel hassled by and fearful of them? The answer will reveal how accepting they will be of you and how easy to work with.

Keep in mind from the time of submission, inventors, along with their products, are being evaluated. Companies consider these questions:

- Is the inventor acting in a professional manner?
- Has the inventor put the necessary effort into developing their product?
- Is the inventor credible? Are they open to learning?
- Does the inventor have reasonable royalty and sales expectations?
- Does the inventor possess the ability to help with ongoing development?

In my position as open innovation director, I spend a great deal of time working with inventors from all around the world. I examine product submissions for ingenuity, functionality, consumer benefit, market position, and cost to manufacture. At the same time, I am evaluating you, and how much effort has gone into developing your product. I usually know within minutes of reading a submission questionnaire if you're ready for prime time. It is essential for me to know if you have completed the proper amount of preparation and diligence. How you present your product is crucial to the evaluation process because I spend a considerable amount of time with division heads, legal teams, and the product development departments as the liaison between company and inventor to ensure the process stays on track. If I think there may be a potential problem down the road, I need to identify it early.

HONEST TIPS

Sleep on It

It's easy to review either terrific products or terrible ones. I have developed an internal code that I utilize; however, when reviewing products that I'm on the fence about advancing, I sleep on it. If I wake up the next day and review everything from scratch and do not feel as strongly about the product as my initial excitement, it's not a good sign. On the other hand, great products always ramp up my excitement level with more consideration. The same holds true when I present a product to company management, and the response is lukewarm or vanilla. You can feel it when you're onto something special. It sticks in your gut. So, if you are up in the air about an idea's prospects, don't force it. Go to sleep and see how you feel in the morning.

Companies are jaded after years of working with half-baked submissions and the occasional oddball. As an inventor, it's essential to change this negative stereotype. When submitting, keep your answers short, descriptive, and professional. Don't try to close the whole deal during the first submission step. A more manageable goal is to create enough interest to keep the conversation flowing and the review process moving ahead to the next level, which will include more technical inspection, such as prototypes and intellectual property. The best advice is to treat the product review process like an interview for a job. Listen, be cordial, and don't do all the talking.

Two Initial Questions I Always Ask

These are two questions that I always ask early in my submission review process, assuming I am intrigued by the product and feel the inventor is on the right track.

- Do you have a functioning prototype?
- Have you filed for proprietary intellectual property protection?

Companies want to know upfront if the product will function under real-world conditions. This is why a functional prototype is necessary. Companies are always curious if the major engineering hurdles have been technically identified and addressed—information that cannot be derived from a sketch, CAD drawing, or sell sheet. It amazes me how many inventors try to submit products when a functioning prototype has not been attempted. I think this is sometimes because inventors are taught not to develop a prototype. They become convinced this route can save money. If you really think about it, they are required to pay about the same amount of money to their "teachers" instead of investing in a prototype. The inside reality is, not having a working prototype sends the wrong signal, right from the start. Your prototype doesn't have to be perfect or look beautiful. It does need to prove concept and function properly.

Even if you do generate early interest in your product from a simple CAD drawing or sell sheet, you will next be asked to provide a prototype, so best be prepared. When there is initial interest by a company in the product, and the prototype question comes up, a "no" answer can bring the review process to an abrupt halt. Why? Because I have seen over the years, companies moving ahead with licensing a product only to discover later that product can't be built. Companies cannot afford these types of mistakes. It's the reason most manufacturers have an established process for vetting manufacturability, refining production costs, and identifying materials. Companies garner details on every facet of a product early on, and if it can't be made to function at the right price, the company will drop the project.

I have witnessed deals collapse deep into the negotiation and signing process because of the non-functionality of a product. Professionally, it intrigues me when public announcements are made about the signing of licensing agreements when this is only the first step in getting to market. A signed licensing agreement does not determine success. The steps to pursuing success include thorough research, strong prototyping efforts proving function, problem-solving, and professionalism. Then it's up to sales and reorders.

INVENTOR INDUSTRY ROCKSTAR

Tom Gray

Tom is the CEO and co-creator of *Make48*. In this time-capped forty-eight-hour competition, four-person teams build a prototype based on a selective challenge and, with access to tool technicians and makerspace equipment, turn their ideas into reality literally overnight. *Make48* is in its fourth season, airing nationwide on PBS Television. The show is filmed in rotating locations such as The Kansas City Union Train Station and the Smithsonian Museum in Washington, DC. Tom grew up in New Zealand and started working with inventors at a young age in Australia. He moved to the United States in 2008 and continued to license products for his own brands, supplying retailers such as Sherwin Williams, QVC, Walmart, and many specialty stores. If you haven't watched the *Make48* television show, be sure to make time for it. In my opinion, it's the best innovation reality show on the air. *Make48* wonderfully combines elements of the disruptive Maker Movement with the spirit of traditional inventors. Tom and his wife, Amy, have done a terrific job mentoring young innovators and building an educational network, without charging them.

Marketplace Vetting

Some other questions I think of while reading submission questionnaires are:

- Can it be manufactured at the right price?
- Has the inventor vetted the marketplace to see if there is anything else like it?

I don't expect the inventor to know the exact cost of production, but it is always impressive when they have a general idea. This might be based upon researching other products of similar size and materials or checking online for costing resources. Often, I receive submissions where a quick check on Amazon reveals the exact product already exists. It puzzles me why an inventor wouldn't take one minute to research there. In many instances, this could save them a great deal of time and energy down the road. Other questions also arise, such as:

- Is the inventor committed to this specific product, or are they bouncing from idea to idea to see what sticks?

- Does the product fit within the assortment of products my company produces?
- Has the inventor identified a real consumer problem that enough people encounter to ensure financial success?

You might be amazed at how fast I filter out submissions that cannot positively answer these basic questions.

Inventor Credibility

I can learn a great deal about you as an inventor and person in a short period of time when reviewing a product submission. I can tell almost immediately if you're new to the licensing arena. I can tell if you are being coached or if you are a seasoned veteran. My initial impression of the inventor continues throughout the review process to license agreement signing. If I perceive an inventor is not ready for prime time or possesses limited product development abilities, it raises red flags. If I have any doubts about your credibility, it does not bode well. I have learned over the years that if the person is off base in the beginning, an agreement is almost never signed. This is not something to be afraid of, simply be nice and truthful.

Meeting in Person

Many inventors want to meet with me in person to show their products. If I had a dollar for every person who wants to fly in from around the country on day one, I'd make a generous living. Be patient and know there will be a time to meet in person after a formal review process has been followed, and your product is worthy of advancing. Approach your business professionally and respect the review process. Provide what the company requires initially and leave the jet-setting for down the road.

When Does the Evaluation Clock Start Ticking?

When submitting your product for review, the evaluation clock starts ticking immediately, for both you and your product. If you are not quite ready to submit, then hold back until you are. I have reviewed well over 100,000 outside product submissions, and there is not a great deal you are going to sneak past me. There are no lucky breaks or shortcuts, so take your time and do things right.

Professional Presentation Etiquette

Approaching companies and pitching your product involves professionalism. It helps if you are concise, knowledgeable, and open to critique or advice. Never lose your patience or temper. Never come across as a know-it-all. Ask the right questions, be polite and understanding. Be quick and to the point with your pitch, and most of all, listen to what is said and follow any advice provided. When you present, keep in mind that you personally are being evaluated. Not only is your product being evaluated, you are also being judged as a professional. If you start on the wrong foot, a licensing agreement may never be reached because they may bail out early. Train yourself to be polite and confident, but not overly aggressive. Handle yourself with professional etiquette. Even if your first product doesn't advance, think of maintaining a relationship for the future.

HONEST TIPS

Product Submission Etiquette

Treat the product submission process in many ways, like a job interview. Here are two tips from the famous book, *What Color Is Your Parachute?* by Richard N. Bolles: Prepare your own agenda by listing questions and curiosities and, when answering a question, speak briefly and allow the interviewer an equal time to respond. I might add, don't try to close the sale in one conversation. Your initial goal should be advancing your product submission to the next level.

A serious inventor needs to listen and converse politely. Language is never an issue for me. It doesn't really matter if English is not your first language. If the conversation is professional, I will do everything to help and understand you. I enjoy meeting inventors who are passionate about their work. The better prepared and presented, the more I want to know about the product. If you approach me with unrealistic expectations, demands of high royalties or advances, or demand an immediate answer, I will quickly shut the review process down. Remember, one of my jobs is to rationally protect the company and its interests. Companies receive an abundance of submissions and can't afford to get sidetracked by a crazy inventor. My role ensures this doesn't happen and, in part, is what I am there for.

If You Can Find a Better Deal, Take It!

If you believe you can get a better deal that results in more sales, or quicker answers, I always say go for it. I understand that inventors need to do what's right for themselves and their families. Yet, sometimes, I receive unreasonable threats from inventors about bringing their products elsewhere if they don't get the answer they want within days. In these situations, I politely cut ties immediately and wish them well. However, I'm never surprised when I hear back the next day with lowered demands and a more rational approach. It happens every time. I do appreciate those moments, but in the end, losing your cool never helps the cause. Just stay rational.

I pride myself on helping a significant number of well-prepared, polite inventors by providing them with focused advice and guidance. However, I do travel a lot and attend many trade shows and industry events. My time is limited, so I use it efficiently to stay on top of the many submissions, emails, calls, texts, and social media outreaches that I receive every day. When time allows, I do my best to walk people through the challenges they are facing. The best way to contact me is through email. I answer all inquiries personally and will get back to you in a reasonable amount of time. (Note: this may mean several days unless I am out of the country, which will take longer.) If there is significant interest in pursuing a licensing agreement, no worries, we will be speaking often!

Not being professionally prepared to submit your product speaks volumes. The idea of submitting multiple, half-baked products randomly to see which one sticks is the wrong route to take. Also, remember, the companies I work with view products that simply combine two different functions into one product negatively, even though some coaches promote this. When I view a submission that is not prepared, or has little understanding of the field, it raises a red flag.

8

Inside Legal Licensing Agreements

Signing a licensing agreement is an important step for an inventor, one that you can certainly be proud of, but only the beginning of your journey to making a profit. Think of the day you sign a licensing agreement in the same way as the day you graduated from high school or college. It's certainly a day to celebrate, but a milestone that doesn't in and of itself ensure success. That's why they call it "commencement" or "beginning." There is still a long way to go.

Many licensing agreements can average ten to fifteen pages, though not all are the same. Besides addressing the core issues of royalty rates, terms, territories, and advance payments, most are filled with boilerplate language addressing standard business issues. When signing a licensing agreement, I always recommend you hire a business attorney to represent your best interests, and not rely on the same patent attorney who filed your intellectual property with the USPTO. They each have different skillsets.

Though I am not an attorney, I have been involved with many licensing negotiations. My closure rate has been high, primarily because of my unique perspective of seeing every issue from all sides and maintaining the trust of both parties. I am well positioned to translate the major concerns of each side and feel it's my responsibility to advise inventors on topics that companies consider important. I also translate on behalf of the inventor topics critical to their interests. In the end, however, each party must make the decisions that are right for them.

What Does a Good Licensing Agreement Look Like?

Licensing agreements revolve around royalties, minimums, advances, exclusivity, geographic territories, and what I like to refer to as industry silos.

Inside Royalty Rates

Royalties, the most asked-about portion of licensing agreements, are an agreed-upon dollar or percentage that is paid by the licensee (the company) to the licensor (the inventor) on every unit the licensee sells. The more units sold, the bigger the royalty checks, which are typically paid out quarterly. For consumer products, the acceptable range for a royalty percentage is typically between 2 and 7 percent of the net wholesale cost, the dollar amount at which the company will sell your product to retailers. Most agreements that I have been involved with, the range is 3 to 5 percent. If the product is not patented or the licensing agreement is not exclusive, the rate will be lower, often in the 2 percent range. I have rarely seen royalty rates over 5 percent: Although rates of 6 and 7 percent can happen, be realistic about what you are expecting. One of your jobs is to gain an understanding of the profit margin requirements within the industry you are trying to exploit, and how this will affect negotiations.

Not being aware of traditional royalty parameters indicates an inventor does not have experience in licensing negotiations, which will not help your overall mission. One mistake you do not want to make is demanding an unreasonable royalty rate without any understanding of what the industry challenges are. When this occurs, the company may bail out on the spot. I have seen this happen more than once. If a company senses at any point they cannot, in good faith, get a deal done because of an unprepared or irrational inventor, or an overly aggressive attorney, they will cut their losses and move onto other projects. During negotiations, if you have any questions about the parameters of the deal or other legitimate concerns, ask upfront to avoid problems. It's always better to ask questions than to come across as unreasonable.

It is the responsibility of the company to provide the licensor with a report detailing accounts where sales originate. Most licensing agreements include language that allows the licensor to audit sales numbers upon request, usually at their own expense, and stiff penalties may be assessed if any discrepancies are found. From my experience, and I have never seen this first-hand, it would be rare that any reputable company would ever institutionally falsify their books to reduce royalty payments. Why? Because companies consider a royalty payment another fixed cost of goods. If the margins are working for them, it is viewed as a reasonable price for success. And if, for any reason, the margins stop working for them, such as a dramatic rise in manufacturing costs or retail competition, the typical remedy is to renegotiate the royalty.

INVENTOR INDUSTRY ROCKSTAR

Inventor Advocate Adrian Pelkus

Adrian Pelkus is a multiple award-winning new product developer with over thirty-five years' experience creating over three hundred different electronic products and high technology processes. As founder, CEO, COO, and CTO, Adrian has helped start several technology companies and is named as inventor on seventeen issued US patents. Adrian reorganized the San Diego Inventors Forum, a 501(c)(3) non-profit local inventors club, to help startup entrepreneurs. A past Entrepreneur in Residence at CONNECT, Adrian also served on the Board of Directors of both the United Inventors Association and US Inventor, organizations dedicated to assisting and educating inventors about their rights. Adrian, who currently advises and mentors inventors through Inventing with Adrian, LLC, has donated a great deal of charitable time and knowledge to helping inventors over many years.

Fair Royalty Rates

The companies I work with do their best to be fair when negotiating royalties. Royalty payments can dramatically raise the cost of goods, and thus the retail selling price. Companies need to reach a four to five times wholesale to cost ratio, and the retail selling price is often double that amount or more. A 1 percent increase in royalty rate can increase the retail to consumers by 10 percent or more. When a royalty rate percentage is too high, the product runs the risk of no longer selling in the numbers projected, which means cutting back on orders. As a rule of thumb, a product will sell 67 percent fewer units at $29.99 than it will at $19.99. Since it's in your best interest to collect royalties on more units sold, root for keeping the retail price down.

Flat-Rate Royalty

There is another approach to negotiating a royalty agreement, suggesting a fixed dollar amount or fraction of a dollar amount, such as 50 cents per unit, for example. Companies may or may not be willing to go this route, but it's worth a try if the suggested amount is reasonable, assuming the recommended amount still fits within the cost of goods. The best time to suggest this approach is when the retail price point is low, and a percentage of royalty would mean pennies per unit. Sometimes, at these lower price points, a company can be flexible. A fixed royalty takes any ambiguity out of the definition of net wholesale. Companies must have a strong desire to get your product to market to pursue this course.

Advances

Many inventors think they should receive an upfront payment at signing separate from the future royalties. Such a non-refundable, upfront payment, in the consumer goods arena, is unlikely. What you might expect is an advance against future royalties. The same goes for other industries, such as authors of books. Though you may acquire an advance, remember any serious income will derive from your product selling well. To accomplish this, the product needs to be manufactured, packaged, and fulfilled at the right price. I have seen advances paid from $2,500 to $25,000 depending upon the industry and projected sales strength of the product. It's not unusual for an inexpensive consumer goods product advance to be around $5,000. Sometimes you can tie your advance into converting your provisional patent application (PPA) to a non-provisional patent filing. Instead of focusing solely on advances, an inventor would be better served establishing a realistic timeline for getting your product to market. The sooner manufacturing begins, the quicker the product will get its first chance to sell in the marketplace, and sustained royalties have an opportunity to begin. That's where the real profit potential is.

Minimums

Minimum sales requirements are generally established to ensure a company's exclusive rights to exploit the product, which are, of course, tied directly into securing an inventor's technology. The uncertainty and risk in precisely predicting future sales without any marketplace sales data usually concerns companies, but they are generally open to establishing minimums. One mistake inventors often make is confusing minimums with sales projections, which happens when they multiply the minimum unit quantity times the agreed-upon royalty, and the number seems low, leading to the natural feeling the company should be selling more. Unfortunately, this misses a major point. Minimums are designed to ensure that a company puts its best effort into producing the product and selling as many as possible, not predicting an exact number of sales. One important aspect to understand is that it takes the same effort to sell twenty-five thousand units of a product as it does hundreds of thousands. Why? Because it takes the same tooling, packaging, and sales effort to get the product placed regardless of actual sell-through. How well the product sells and is reordered remains up to consumer demand.

How the product performs in the market when, hopefully, selling through will ultimately determine the size of your royalty checks. A top priority is for companies to leverage their retail customers into ordering and displaying the product. Estimating minimums for a product that has never before entered the marketplace might result in actual sales falling short of projections. Many things outside the company's control could adversely affect sales, such as delayed production and postponed orders, so companies aren't going to risk investing lots of money and unreasonably losing the exclusive rights to the product. Most licensing agreements do not include any minimums in the first twelve to eighteen months. Typically, the minimum sales numbers increase over a span of two to three years as the product eventually rolls out. For example, for a twenty-dollar retail item in the second year, minimums could rise to twenty-five thousand units and the third-year bump up to fifty thousand units. These numbers are not atypical. The DRTV business is a little different, and we'll cover that industry later in the book.

Territories

The territory in a licensing agreement is another major component. Most US companies are interested primarily in North America. Sometimes the territory can be worldwide, but you will need to determine if the company can even service the world through its own subsidiaries and affiliations. Keep in mind that if your product is shipping worldwide, you will need to have filed for worldwide intellectual property rights, sometimes accomplished through a PCT filing, or at the very least, remain within the international patent filing period so the company might consider filing on your behalf.

Indemnification

There is typically an essential topic in most contracts called indemnification. As companies perform their due diligence, they are sensitive to discovering anything hidden. For example, if a patent was not filed correctly in good faith, or the inventor was not forthright about their corporate ownership interests or assignments. Therefore, the company will understandably insist upon legal language protecting them before signing in case any illegitimate claim was made, making the inventor liable for their own errors. A reasonable attorney needs to ensure the language is clear and does not put an inventor into a liability bind they cannot afford. I've never seen a company drop the indemnity clause, but often they will grudgingly make it reciprocal.

Industry Categories or Silos

Sometimes an inventor's technology may be utilized in different industries. Many companies do not mind negotiating an exclusive licensing agreement in just their field, allowing the inventor to shop their technology in other areas with non-competitors. I call these different silos. This offers the opportunity to sign separate agreements in the two different industries.

HONEST TIPS

Industry Silos

Silos, as in feed storage towers on farms, refers in the world of open innovation to something more akin to tech transfer. Some products provide crossover solutions in different market sectors, which may create additional licensing opportunities acceptable to multiple parties in different industries. For example, the consumer housewares industry is entirely different from the commercial restaurant supply and equipment fields, and an inventor's product might work well in both industries.

Exclusivity

An important licensing concept is exclusivity. This is an integral part of licensing agreements, determining the company's sole right to manufacture and distribute the product. Sometimes, companies will allow the inventor to service non-competing specialty accounts, or consumers directly, or even operate from their website, if it does not interfere with the company's relationships with their large volume accounts. Otherwise, companies typically want an exclusive relationship, so it's up to you to carve out any niches you feel are important.

Discounts

Royalty checks may be negatively affected by legitimate discounts, damages, and returns. Discounts may arise from incentives the company provides its retailers to increase sales, such as volume rebates or advertising allowances. Note that royalties are always paid on the "net" cost of goods number. Damages are what they sound like; defective products returned to the supplier. Retailers often negotiate a damage allowance upfront in the agreement and simply discard the broken goods. A return of product occurs when a retailer has too much stock on hand and needs to balance their current inventory with other products being offered by the supplying company. Some companies hold back a small reserve amount anticipating future returns. Royalties are paid on goods the company actually sells, based upon generally accepted accounting procedures.

Non-Binding Term Sheets

The companies that I work with have all developed non-binding term sheets that generally address the significant legal issues described so far, including royalties, advances, minimums, and territories, before the attorneys get involved. Although the term sheet may not be legally binding, it allows the parties to pre-negotiate noteworthy issues in advance, so there are no major surprises when the attorneys get to negotiating the final details. All this saves both parties time and money while serving as a reality check. After all, if the two sides cannot quickly agree on the significant parameters, why bother moving ahead? Remember, every licensing agreement is different. I have partaken in agreements that have actually allowed independent inventors to purchase their own products directly from the same factory, ensuring lower pricing that the company has already negotiated. This is sometimes called piggybacking and generously allows the inventor to leverage the company's purchasing power. Though many licensing fundamentals cannot bend, there are many ways to approach unique circumstances, so if it's important, ask. You must know where to look and intelligently present your case, making sure it does not inadvertently harm the company. This requires rational discussion, goodwill, trust, and communication. Anything that can be worked through in the early, non-binding process will save both sides unnecessary legal bills.

Ownership of Intellectual Property

If you are offered a licensing deal, make sure that the agreement ensures you are the owner of the intellectual property, and you either have the rights to any improvements on your intellectual property, or you are paid a full royalty on any product that uses your original intellectual property, regardless if it is paired with any other company intellectual property. Expect the possibility of the company adding improvements when preparing the product for market. Assuming a company is willing to agree to this arrangement, you should feel confident that you are dealing with a quality company. Also, note that the company will require you to formally assign your rights to the patent when the agreement is signed. If you have a co-inventor or any other legal signatory issues, they need to be addressed before closing.

Even though the cost of goods is essential in finalizing licensing negotiations, the quality of your intellectual property is also important during negotiations. The stronger and more defensible the patent, the higher the retail price point a product will support, meaning a higher royalty. Protectable intellectual property helps all parties by enhancing the company's proprietary position in the marketplace and increasing margins. Some naysayers may dismiss the value of patents these days, but strong IP will still earn you a higher royalty percentage. Inexperienced inventors sometimes look at the retail selling price and compare it to their royalty per unit and think they are receiving too small an amount of the total, while the company is getting rich. Remember, the company is taking on all the risks, including the cost of inventory, fulfillment, and other required expenses, which greatly reduces the company's bottom line. In many ways, the company also serves as your angel investor.

Attorneys and Negotiations

I am often amazed when novice inventors and inexperienced lawyers try to carve up contract language during negotiations. When a hand is overplayed, and I have seen this many times, it quickly reveals a lack of experience in licensing matters, which raises questions as to credibility. Inexperience and ignorance can make an agreement difficult to reach, even preventing execution and signing. Once again, the inventor is being evaluated from the start, just like the product submission itself. Your professionalism and knowledge in all legal matters are also being assessed. As for your attorneys, never forget, if you actually want to get a deal done, they report to you, not vice versa. This is something worth remembering in life as well: No matter the negotiation, it's your life, not theirs.

A word to the wise, use a good business attorney to negotiate and secure your licensing agreement, not your patent attorney. Patent and business attorneys represent two different areas of expertise. I have seen terrific patent attorneys destroy business deals by not understanding basic terms and definitions. I also caution against using a high-powered, expensive corporate attorney. Even if they are a friend, these lawyers often think they are representing a Fortune 500 company and send back first draft agreements in a sea of red ink. Without fail, this causes immense delays and sometimes even brings negotiations to an end. It's a simple fact that a large company has more options than you do. If you negotiate like they have something to lose by not signing with you, you might be in for a rude awakening. I suggest saving your ammo for when you've done a few deals and have plenty of money.

The reality is that most inventors are ordinary individuals; many are going through the licensing process for the first time. They're pursuing licensing with products that have no sales history, clout, or leverage in the marketplace. You have a personal responsibility to make sure that your attorney is representing you and not serving their own ego, which can ruin a deal. Once an agreement gets shelved or tossed, it's difficult to get things started again. Keep in mind that most large companies also have big-ego attorneys, so, throughout the process, stay above the politics and fray. Do your best to keep both sides focused on the product and making money together and use what you learn from the process to help negotiate your next agreement.

Disadvantages of Licensing

One of the significant disadvantages of licensing your product is that you will not earn as much money per unit sold. Another disadvantage: In many instances, you will not have control over the design, packaging, and or launch details, which include where your product is manufactured or sold. As the inventor, your ideas are valued; however, the company will make all final decisions.

Another disadvantage of licensing is settling on the wrong company. Often, inventors decide on a company not large enough to command enough marketplace clout to earn substantial money. I sometimes witness licensing deals announced by marketers or inventors on social media with companies that I've never heard of before, even in my arenas of expertise. If I haven't heard of a company, there is a good chance they are not leaders in the industry and lack a stable of household brands. In these situations, I worry about the inventor not being able to reach their income potential with potentially under-resourced companies. I recognize the social media announcement of a licensing deal being signed is primarily an opportunity for a marketer to shine the spotlight on their own company, but when I see these announcements I always professionally think, let's see in two years how many units were sold, and if the inventor earned back more money than the marketer charged them upfront.

The Time it Takes from Signing to Shelf

A question that I am often asked is how long, after a licensing agreement is signed, does it take for a product to get to market. New products that launch can take up to a year to begin rolling out as traditional companies need to complete development work, set up tooling, design packaging, manufacture inventory, and so forth. Specialty stores and nimble larger retailers are the first to launch. Many of the big box stores have shelf plans which take a great deal of time and lots of meetings to access, so it might be closer to two years before you get into all the stores across the country. You generally need to be patient as sales ramp up. Hopefully, on the back end, your product will have a long shelf life.

9

Going to Market (Part 2)

The Thirty Steps of Going to Market Directly

The alternative to licensing your product is going directly to market on your own. Understand upfront, though the profit potential is potentially greater, the risks are also exponentially higher. There are certainly other positives to going direct, including being your own boss, making your own decisions on development, design, and marketing approaches, not to mention deriving satisfaction from creating your own business and generating personal acclaim. You need to think through all the components and investment dollars that will be necessary, however—including many challenges that may ultimately be beyond your control. In this chapter, I will identify these challenges. As you'll see, the list is long. For full disclosure, I have participated in taking a product directly to market several times and have experienced both exceptional and depressing results.

■ ■ ■

Selling Directly to Consumers

There is more opportunity today than ever before of bypassing the traditional retail stores and selling your product directly to consumers. Producing smaller quantities, marketing through social media on the internet, and driving customers to your website can quickly get you up and running with national, or even international, exposure. This can validate whether you are producing a product that can generate reorders and become sustainable. Selling directly to consumers will also enable you to bypass big retailer credit checks, fulfillment challenges, and large inventory commitments. There are also resellers like Amazon that make it easier to connect directly with consumers.

Marketing Plan

It's critical, before launching your product, to identify the market you want to conquer and develop a strong marketing plan to accomplish your goals. How you get the word out, particularly without an identifiable brand, will be critical to establishing early sales and determining if there is a reason for further financial investment. If no one knows about your product or its benefits, it will be difficult to gauge sales. Using crowdfunding platforms such as Kickstarter and Indiegogo might be one approach. This typically requires a great deal of social media support on Facebook and Instagram, so you need to budget in the cost of such a campaign. Of course, you can also pursue the social media route without a crowdfunding strategy and drive customers directly to your website.

There are also many online retail opportunities these days, such as Amazon, eBay, Etsy, or The Grommet that allow independent product developers to access millions of potential customers easily.

One challenge is that your product will be simultaneously exposed to competitors and would-be infringers who closely monitor these sites for hit products. When a new product is selling well in the marketplace, many others soon take notice, and the door opens to copycats. So it is critical to have a marketing plan in place to drive sales, in addition to strong patent protection and a trusted manufacturing partner.

Never forget that competition represents a significant challenge when selling a product. First, you should search thoroughly to see what else is out there already. Even if nothing exists and you are successful, however, it won't take long for similar products to pop up. Sometimes they are legitimate alternatives with new features, and sometimes they are direct knockoffs. As someone said to me once when launching a new product: "We're not in school anymore." It's America, and companies that have been in an industry a long time are not going to lie down and surrender a category. Therefore, your marketing plan is critical. Unless you are just fooling around and happy selling a few pieces, you need to think through getting out of the gate quickly and creating a beachhead for future success, before others swoop in.

Retail Selling Price

If you are going to be competitive, while protecting yourself from others undercutting you, it's imperative that you get your retail price point in line with what consumers will spend. Never forget, the customer determines the price point. There is no bigger disappointment than a good product launching at unaffordable high retail, and no one ends up buying it. If you are not sure where the retail price point needs to be, search for similar products online and ask salespeople at retail stores. Once you have established a saleable retail, you need to work backward and determine the cost of goods, ensuring your profit margins are reasonable. To arrive at the appropriate cost of goods, you need to establish the physical portions of the product, including claims, mechanics, and materials. It's yet another reason, beyond filing for a utility patent, why building a functional prototype matters. This isn't a game any longer. You need to be ready.

Going to Market Through Traditional Retailers

Selling a product directly to large retailers is a difficult challenge today. First, getting through the buyer-review process can be difficult. Why? Even if you know whom to contact and approach inside a big retailer if you offer only one SKU (stock keeping unit) it is virtually impossible to get anything going initially beyond a small test order. Retailers want fewer vendors, not more. Without an assortment of other products to balance inventory, they don't want to put themselves at financial risk carrying large quantities they may have to potentially markdown and liquidate later. In the beginning, large retailers have no idea of an individual's or small business's financial ability to support the enormous responsibility of stocking their shelves and warehouse. Even if you can produce the opening order, the question remains: Do you have the financial ability to ramp up inventory if sales suddenly take off? Do you have the factories in place to increase production? What if the retailer runs an ad, can you ramp up and ship on time? The results can be disastrous for big retailers if you, the vendor, can't financially and mechanically handle the demands of success. It's hard to build trust overnight if you are new to the game.

Utilizing Distributors

If you are a single SKU company, another avenue to explore helping you fulfill your products is a third-party distributor, which can alleviate a retailer's concerns about your financial resources. It is important to locate the right distributor, so listen to guidance from the merchants at your chosen retailer: Most retailers have distributors they trust. Throughout your journey, you may end up using more than one distributor for different retail stores. While you're vetting distributor candidates, take advantage of sales representative connections who want to sell your product. A lot of times these folks can be met at industry trade shows. Remember though, distributors can charge fees ranging between 10 to 15 percent of the wholesale price to fulfill and distribute your products.

Independent Sales Reps

Another cost that you will need to account for if you sell to retail stores, typically around 5 percent of the wholesale cost, is for independent sales representatives. While large companies typically have their own in-house sales teams, smaller ones use independent reps who may sell many different product lines within an industry. Sometimes you may find a national sales group that will cover the entire country, or you may want to choose independents by separate regions.

The Thirty Steps to Market Checklist

Taking your product directly to market by yourself is riddled with many challenges. You need to assess the best route, either licensing, direct to consumer, or directly to the retailer. I put this list together many years ago at a speaking engagement when someone innocently asked me the basic steps involved going directly to market. They were expecting a brief response, but I got to thinking on the spot and answered in full. A half-hour and twenty years later, this list still holds up today.

1. **Idea Generation.** Whether you license or take your product directly to market, the first step in the development process begins with generating a new product idea that solves a big enough problem for enough people, which ultimately has the potential to generate significant sales.
2. **Marketplace Vetting.** You must thoroughly research the marketplace and identify the industry you plan to exploit by going online, visiting retail stores, visiting trade shows, and asking questions of in-the-field experts.
3. **Early Product Development.** Translate your initial concept into mechanical design drawings. If you are not an engineer, you might require outside help with this step.
4. **Prototyping.** It is important early in the process to build a working prototype that proves function. This will also form the basis for the physical claims that eventually form the basis for your patent filing. The prototype does not have to be attractive, but it does have to function!
5. **Engineering Plans and CADs.** Finalize your engineering plans in CAD format, which can be used to cost goods and determine if the product is manufacturable.
6. **Patent Search and Filing.** You may choose to use a professional intellectual property attorney at this point to search when filing for a utility patent. The first step is a provisional patent application (PPA). After twelve months,

it must be converted to a non-provisional patent filing. Decide within the first year if you want to file internationally with a worldwide PCT filing.

7. **Manufacturing Facility Selection.** Locating a factory and developing a manufacturing partnership is the next step. Establishing you have the technical and financial resources to sustain your project will be critical, as well as vetting the factory's capacity, tooling, and production costs and experience. You will need to explore several facilities to select the one best for you. Whether in the United States or overseas, you may want to visit these factories or use a broker you trust.
8. **Payment Terms and Upfront Deposit.** Most factories do not extend credit to first-timers. After a relationship is established over time, you may be able to amortize future tooling and production costs, which is an accounting term that refers to "allocating the cost of an intangible asset over a period of time." In the meantime, you need to budget early payments upfront.
9. **Manufacturability Analysis and Design Tweaking.** Once you have your finances in order and a manufacturer is chosen, you will need to work with their engineers and design team through prototypes and CAD drawings. This is to ensure whether the product is manufacturable or needs to be further tweaked. And this is where a thorough understanding of engineering details can be helpful.
10. **Raw Material Decisions.** Modern CAD simulators and software programs allow you and the engineers to sample different materials for optimum function and design,

saving time and money. Materials can include anything from metal, plastic, rubber, foam, wood, cloth, and more.

11. **Final Costing.** Once you have established specific materials, and the final design, the cost of materials and production must be determined. Note that larger quantities will bring costs down, but initially, without actual sales results, you need to be conservative on your estimates.
12. **Tooling.** This is an essential part of the process where the manufacturer assembles all the components needed to fabricate your product. Components used in tooling include fixtures, jigs, molds, gauges, dies, patterns, and cutting equipment technology. Learn all that you can about these terms and what they do, or hire an expert in the manufacturing field.
13. **First Off-Tool Samples and Quality Testing.** Once your product is tooled up, there will need to be a first off-tool sample run, another critical part of the production process. A small run of goods is necessary to ensure proper function, work through any minor tweaks, and determine if your product is ready to move forward to mass production.
14. **Full Production Run.** Next is a full production run. Everything from final production costs to quantity on the first run must be finalized. Manufacturing facilities have production schedules, and you need to know where your product fits within their timeframe, which might delay you from a few months to a year. Be sure to get everything in writing along the way.
15. **Packaging and Instructions.** Product packaging is critical, and you must have specific instructions and guidelines included. Note that the average time a customer spends initially taking in your product on the selling floor is only

1.9 seconds, so you must put your best foot forward. Besides highlighting selling information, most packaging has special labeling including warnings, hazards, safety materials, and more.

INVENTOR INDUSTRY ROCKSTAR

Target Fundraising Expert Kedma Ough

Too often, great ideas fail to see the light of day because the inventor doesn't know how to secure the necessary funding throughout the inventing process. No matter what your idea might be, there is funding available to build a solid business or invention around it. Kedma Ough, MBA is a recognized alternative funding thought leader and has advised more than 10,000 inventors and entrepreneurs on a wide range of financing options. Over the last twenty years she has held leadership roles with both the SBA Small Business Development Centers heading up Innovation and the Women's Business Centers. Currently, Ough is the VP of business coaching for Conquer, a company overseeing advising and guidance for hundreds of small businesses servicing the home services industry. In 2019, Kedma's terrific book *Target Funding* launched: a navigation funding roadmap to help inventors and entrepreneurs locate grants, funds, and resources to scale their ideas and reach their dreams. Kedma has been a long-time Board Member of the United Inventors Association.

16. **Initial Quality Control and Inspections.** When your product is ready to ship, it must undergo initial quality inspections. Everything must be 100 percent functional without damage before shipping. If you are dealing with

manufacturers overseas, you must pay a local inspector to test a random sampling and supply you with an official report.

17. **Final Payment for Goods.** Once the final packaged product is tested and ready to ship, payment needs to be made, which can be concluded via bank wire transfer. In time, as your credibility and relationship grow, factories may extend you a line of credit, or you may even work off a letter of credit with the retailers you are supplying that can tie into your bank at home, assuming such special terms are negotiated upfront.
18. **Shipping to the United States.** When manufacturing overseas, it can take weeks for products to reach American shores. Then the shipment must be examined, processed through customs, and loaded onto trucks or trains. All of this takes time and additional expense.
19. **Duties.** Coming into America with new products is never easy. The regulations are long and complicated. The formulas used to determine product categories and the cost are highly complex. Different categories carry different percentages that you have to pay, using formulas. A select few products are not allowed into America at all. You must know your product category rate and factor it into the cost.
20. **Port of Entry Fees.** At US customs, there is a port of entry fee, which is charged for processing, inspections, and passage through the review process. These fees also include transferring any cargo containers from ship to truck. Port of entry fees include all taxes and costs associated with US Customs and Border Protection (CBP). If you need help with this, contact a port of entry specialist to find out the taxes and fees for your specific product category.

21. **US Shipping via Truck or Rail.** Whether a product is manufactured overseas or in America, the product still must be eventually shipped between docks or warehouses by rail or truck. The actual cost will depend upon carrier, quantity, weight, and distance.
22. **Fulfillment Center Selection and Warehousing.** You will need to contract with a fulfillment center to warehouse your goods. They can also help take orders, pick goods, box, palette, label, ship, and even bill retail stores and individual customer accounts. Retailers today use highly automated receiving centers that rely on all shipping boxes being labeled correctly and using the latest computerized ticket system. Mistakes will lead to very costly chargebacks and penalties.
23. **Sales.** Once your product is ready, you will need to sell it, which will require an effective sales team or independent sales representative. There is, of course, a percentage cost to this. The sales team will negotiate the terms each retailer requires, including the cost of goods, suggested retails, advertising allowances, damages, and returns. Note that some retailers will also have their own quality assurance tests, and all will require product liability insurance.
24. **Order Taking and Processing.** You will need to contract with a call center to receive phone and internet orders. There is a specialized software designed for warehouse processing and fulfillment.
25. **Fulfillment Order Picking.** Orders are accepted and pick tickets generated by a computer. Warehouse staff receives the orders taken by the call center, processes them, and then ships. There are many order fulfillment software choices, including inventory tracking capabilities.

26. **Shipment to Retail Stores and Customers.** The orders are placed back in containers on eighteen-wheeler trucks, which are freight brokered. All this means more fees moving freight to its destination. Accounts can be established with freight forwarding companies, UPS, or FedEx to ship products to retail stores, as well as directly to consumers. You need to follow all the retail shipping guidelines carefully, and all are different.
27. **Billing.** Once your products are shipped out to customers, a system of billing and processing receivables is essential. Having a fully functioning accounts receivable department or outsourcing this service to a dedicated company is necessary to handle invoicing, processing purchase orders, establishing vendor accounts, and managing inevitable returns.
28. **Money Collection.** It's essential for your cash flow requirements, as well as reinvesting in inventory as your company grows, that you collect your money on a timely basis. Direct sales customers pay when the product is delivered using credit cards, although you need to account for processing fees and deductions. Most department stores and big box retailers, however, pay their bills in ninety days, so you need to account for the lag, which can put you at risk financially while waiting for your funds. Another major concern with major retailers is if the product is not shipped on time, or the boxes are improperly marked, they will charge you back hefty fees. This can arise from something as simple as not putting the shipping labels on the proper corner of the box, or improperly identifying the Universal Purchasing Code or UPC. The UPC

allows stores to monitor and control inventory, pricing, and shipping costs. Prepare to register each product to receive a UPC, which involves another cost. Once you develop a relationship with larger retailers, you may be able to provide them with an upfront discount in return for their providing financial letters of credit (LC) and payment upon release of goods from the factory. However, this is not available to most startups.

29. **Inventory Control and Sales Analysis Systems.** In keeping track of your product sales, you will need a high-quality inventory tracking system to monitor your inventory and analyze sales. This is critical to making decisions on manufacturing, inventory, packaging, sales, and shipping.
30. **Reordering.** You will need reorders for your company to survive. Everything up to that point is a dry run. Just like licensing, that's the first time you will know how big a hit you have on your hands. Then you'll need to invest in producing more goods to handle all the new orders.

INVENTOR INDUSTRY ROCKSTAR

American Dreams Dara Trujillo

Dara Trujillo has more than twenty-eight years of retail, theme park (Walt Disney), and TV shopping experience (HSN), creating worldwide concepts and products. While at Disney, she learned to persevere through tough corporate boardrooms to be able to create amazing products and experiences that are still enjoyed today. These experiences are found around the world, including the Bibbidi Bobbidi Boutique, Goofy's Candy Company, expanding the Pin Trading experience, creating Duffy the Disney Bear, and many more. Most recently, Dara was VP of Merchandise Business Development for HSN, where she created and managed the American Dreams Program and Entrepreneur Academy, helping entrepreneurs realize their TV retail dreams. Currently, Dara is the Chief Merchandise Officer for SLC Group Holdings, an angel investment firm led by Sandy Cleary, where the top priority is finding amazing entrepreneurs who want to go to market directly and helping them from startup to marketplace.

In summary, if you're going to take your product directly to market, you must become an expert in every facet of the fulfillment chain. You will also need enough human and financial resources to get off the ground and stay afloat until your business stabilizes. Taking a product to market on your own entails a significant amount of effort and risk that should never be underestimated. Recognize that someday, if, in fact, you realize enough revenue, reorders, and are maybe even generating a small profit, it will be necessary to reinvest your positive cash flow back into manufacturing more inventory to keep the show rolling. Then, of course, the product must continue to sell well, so the challenge of sustained growth continually repeats itself. If this all sounds overwhelming, which would be understandable, strongly consider licensing.

SECTION TWO

Inventors "Beware"

10

Clever Marketers Who Charge Inventors

On the side of every cigarette package sold in the United States is this warning label: "Caution: Cigarette Smoking May Be Hazardous to Your Health." Wouldn't it be nice to require the same type of warning label on every promotional message that invention industry marketers post? Then, it might read, "Caution: Companies and People Who Charge Inventors Upfront May be Hazardous to Your Wealth."

As you can probably tell by now, I don't take kindly to invention industry marketers whose business model is to exploit inexperienced inventors who harbor unrealistic dreams of getting rich, charging them for, quite frankly, mediocre assistance. I have a front-row seat for this occurrence every day. I am usually the one tasked with bringing unprepared inventors back to earth: those people who, from the beginning, never had a chance of making a profit from their new product idea. Aspiring to become a great inventor is laudable, but when you're paying thousands of non-refundable dollars for uncollateralized advice by so-called experts that invariably leads to rejection, well that's where I draw the line. From my experience, it's far wiser to spend your money on tangible product development initiatives, materials, intellectual property protection, consumer research, and focused support services from industry experts, than to count on these intangible marketers for advice or profit.

No doubt, there will always be those entrepreneurs who start businesses with an initial desire to help inventors. They enter the invention marketing space with a variety of motivations. Some are led by passion and entrepreneurial spirit. Others may have experienced some level of new product success and suddenly feel qualified to help others. I personally love assisting inventors, so I understand these sentiments.

The problem is this: I often witness an abrupt change in this mindset once helping inventors becomes a livelihood. Almost without fail, these businesses morph into cash-flow driven, as opposed to profit-driven enterprises. What do I mean by this? When back end, percentage-based revenues generated from legitimate sales success do not materialize, it does not take long before the money necessary to operate these businesses dries up, and good intentions soon wane for them, because it becomes too difficult to find and license enough revenue-producing hit products to generate sustained financial success.

HONEST TIPS

H.R. 1907: The American Inventors Protection Act of 1999

This law requires any invention promoter, before entering into a contract for promotion services, to disclose to a customer in writing: (1) the total number of inventions evaluated by the promoter for commercial potential in the past five years, including the number of positive and of negative evaluations; (2) the total number of customers who have contracted with the promoter in the past five years; (3) the total number of customers known by the promoter to have received a net financial profit as a direct result of the invention promotion services provided; (4) the total number of customers known by the invention promoter to have received license agreements for their inventions as a direct result of such services; and (5) the names and addresses of all previous invention promotion companies with which the promoter or its officers have collectively or individually been affiliated during the last ten years. Sec. 102 establishes a federal cause of action for inventors injured by material false or fraudulent statements or representations, or any omission of material fact, by an invention promoter, or by the promoter's failure to make the required written disclosures.

Every inventor should know the new product launch failure rate is high, not just for others, but for you and your product too. The same goes for marketing companies. They simply can't partner in enough successful products. This is when it becomes clear that to survive, let alone flourish, they need far more pay-up-front clients. It doesn't take long before the initial enthusiasm for helping inventors takes second place to generating the cash-flow necessary to sustain their lifestyle. Remember, they have gone to a great deal of effort to launch their business and have invested significant resources. They'd lose money if they were to drop out now. And this is where it gets interesting because they each take a slightly different approach to generate cash flow, though none consistently deliver bankable marketplace success for inventors, something they do though command for themselves. Some are cleverer than others, so you must look carefully to understand what they are up to fully.

Compared to, let's say, talented authors, musicians, or athletes who have agents helping them sign contracts and earn a legitimate percentage of future income, professional marketers in the invention world convince inexperienced inventors with little knowledge, talent, or skills, to sign up for their "helpful" programs and charge them. What if this were happening in the literary, music, or sports fields? What if agents convinced aspiring authors, musicians, and athletes with no experience to take their special course teaching them how to write, perform, or play? Imagine if the pitch was, "Sign up with us, whether you have any talent or not, pay us and we'll help coach you how to play the piano, write a novel, or play baseball, then we'll point you toward a record label, publishing company, or professional sports team who will no doubt sign you up and you'll be sure to make lots of money." Ummm, good luck. Right?

Hopefully, everyone can see the absurdity of leveraging the limited assets of average inventors, who have no idea what they are doing, particularly the first time around, with the subliminal pitch they are going to become rich. Instead, someone with expertise should explain upfront the odds are awfully long and success rates incredibly low. Someone with actual industry experience should be on the payroll, not generic talkers and company cheerleaders, evaluating their product ideas and honestly explaining to them they are not ready for prime time, and then *not* accept their money. It reminds me of when I was advised as a child, after striking out in Little League on a curveball, "You are never going to make it to the big leagues as a baseball player." I got over it and moved on to lacrosse.

It's just so ridiculously easy to seduce naive inventors into paying money for a dream. In many ways, it's like small-time gamblers unable to resist the temptation of buying lottery tickets. It's like taking candy from a baby. Remember, 78 percent of inventors inherently believe they will become millionaires from their invention. It doesn't take long before the marketers' evolving mindset begins to justify their own existence by asking, "What is so wrong with our charging upfront? Don't we deserve to make money too?" And that's when the full switch is completed, and the business formally evolves into one dedicated to promoting themselves and their egos, along with prioritizing their own cash flow, passive income, and dreams, not yours. Of course, they disguise everything with sympathetic speeches and generic educational posts but peel back all their marketing, and the results are clear: They're making money, and you aren't! If inventing were so profitable, they'd be doing it themselves.

HONEST TIPS

Rather than spend thousands of dollars on inventor coaches, here are other ways to allocate your money:

- Every successful inventor that I have ever helped license a product, who had significant success in the marketplace, developed a working prototype to prove function. They didn't simply mail in a drawing or animated sell sheet.
- The best videos, which are essential these days, are made from functioning prototypes.

- One of the first questions I ask is if you have researched or filed for a patent. If no, it's an early indication I am probably dealing with an amateur who has been steered that way.
- Visit industry trade shows to understand better the industry you are trying to exploit.
- You can easily do your own research finding inventor friendly companies. Go online. Go to retail stores. Go to trade shows. Read trade publications. Attend local inventor club meetings. Search company websites, or ask receptionists for product submission details.
- Remember, if you have a limited budget, how do you want to spend your money best?

There is an old saying, "The money is not found in finding gold, it's in selling the picks and shovels needed to find the gold." That large pot of gold is what every inventor is looking for. And inventor-help companies point up to the hills and say, "The gold's up there!" Then the company, with zero risk, sells the picks and shovels, often at high profit margins. The company is not consumed with whether there is gold in the product because finding it is not really part of their profit model. The only reason they might care about inventor financial success is so they can market that rare win to sign up other paying clients.

The worst part is they tout to the inventor that licensing is simple: Just come up with an idea, start paying them, and prosperity is just around the corner. Turn your ideas into goldmines! You may not even need a functioning prototype or IP protection to sell your idea. What the heck, it's as easy as pie. But it turns out these folks have their own dreams too: to leverage and profit from yours. You do all the work and make the pie; then they eat it. And if you are one of the lucky few who even signs a licensing agreement, they market the good news across every social media platform known to mankind. "Look, this person just signed a licensing deal!" as if subliminally saying, "You can too!" Yet, they always leave out this essential ingredient; signing a licensing deal is just the beginning of the profit process and proves absolutely nothing as to actual marketplace success. Most signed licensing agreements never generate significant royalties. Never forget, signing a licensing agreement is just one step of many along the path to potentially generating profit. It's certainly not a guarantee of financial success. Your real excitement should start the day a reorder comes in. But then what's beautiful is some marketers then leverage an inventor's naive thrill of being featured in a public media post to boast about their good fortune signing a licensing agreement, even though no profit has been generated, turning the inventor's words into an endorsement for them when there was no profit. You must admit, it's downright clever—and scary—and revealing of their true character.

Some of these inventor-help companies dodge the more significant issue of disclosing their actual sales and marketplace success numbers, which are dramatically low. Even worse, I've seen marketing companies promote on social media other successful inventors whom they didn't help or know. They highlight these inventors' success stories to support their own company's subliminal marketing efforts, giving the impression they were somehow involved in these achievements.

HONEST TIPS

When inventor marketing companies charge you up front, it should raise a red flag. These are the questions I would want answers to before signing on:

- Precisely what services to help inventors do you provide? What do you charge for each?
- Do you charge only upfront, or do you partner with inventors and collect revenues from licensing sales and profit success down the road?
- How many paying clients do you currently have?
- How many have you charged since you started your business?
- How many of your total paying clients have signed licensing agreements with your guidance?
- What percentage is this number of your total number of charged clients?
- How many of your clients that have signed a licensing agreement have actually made more money back in royalties than they paid you for your services?
- How many of your total paying clients have earned over $100,000 in royalties?
- How many of your total paying clients have earned over $1 million?
- Can we please audit your numbers? (You know, "trust but verify.")

It would help if you did your homework vetting these inventor marketing entities. When considering an invention-help company or coach, conduct a deeper dive. Here are a couple of essential questions that you should ask upfront. What are the real statistical odds of my signing a licensing agreement? And, how many of the relatively few licensing agreements that you have been involved with, compared to your total client list, actually generated more income for the inventor beyond what they spent on your invention submission or coaching services?

I am the recipient of many gold-seeking inventor submissions every single day. First, many submissions are not ready for prime time, even with supposed coaching (many times, by the way, from assistants far down the pecking order with limited experience and success). When I know an inventor spent significant money to be represented or coached, and they are still not ready for prime time, it literally breaks my heart. The sad part is that the representatives of these companies should know better whether a product has a fighting chance of getting to market, or they should otherwise stick to industries they have real experience in. Unfortunately, all of this leads me to usually be the one to rein in the inventor's dreams, at no charge, while thousands of dollars are paid to the marketers.

Becoming involved in my local Connecticut non-profit inventor organization many years ago allowed me to learn about the struggles that inventors are dealing with and which middlemen companies sought to exploit them. Soon after, I was asked to be a board member of the national 501(c)(3) the United Inventors Association. The UIA was a more substantial national platform where I could hear from more inventors around the country and world. I wanted to know every detail of the industry from the ground up, and more about every inventor challenge.

I soon discovered a variety of companies and individuals that can be described as invention submission companies, consultants, and inventor coaches. Not long after, while signing a licensing deal with an inventor, I came head-to-head with an invention submission company on the West Coast that no longer exists today. The company had made so many mistakes they eventually went out of business. However, before advancing to a licensing agreement, I wanted to know whom I was working with. So, I looked deeper into the company, and their inner workings and what I found was ugly. I found they were offering nonexistent services and charging heaps of money to inventors for additional, unnecessary help and embarrassingly shoddy work. I went further and delved into their records and examined their actual success rates. Their real hard numbers of inventors receiving legitimate licensing deals were minuscule at best, a fraction of 1 percent, and that was just signing licensing agreements, which is only the first step and does not guarantee future success.

Innovation Industry Entry Without Barriers

The nature of the invention industry is there are no real barriers to entry, no formal criteria established for exposing ethical leadership, and no enforced rules or regulations to rein in those entering the inventor-help business. It's like the Wild West. The community often reminds me of the HBO TV show *Deadwood*, where a shifty saloon keeper has every opportunity to become sheriff or mayor. There are no universally recognized degrees or certificates one can earn to become an "expert" in the field. Entering the innovation arena is different from going to law school, studying hard, and passing the bar. It is, in fact, the exact opposite. It is an open platform where anyone, with or without credentials, can step right up, hang their shingle, and start doing business without formal training or any proof of expertise.

When I was a kid, I remember my dad bought a boat. The first day we all hopped in and took off across the water and banged right into another boat. None of us knew how to drive a boat. We had no experience. I never understood, even as a child, why you needed a license to operate a car but didn't need one to operate a boat. We could have all drowned. It was crazy. The same goes for the inventor world. You don't need a license to operate one of these businesses.

Yet, the innovation industry is filled with a never-ending, eclectic group of hopefuls rolling in and out. And not just inventors mind you, but those who are hoping to create a company and charge others too. They seem to show up out of nowhere. Being completely honest, this is how I also originally got started in this arena, so I understand the initial lure, but in my case, there are a few significant differences. First, I learned product selection and retailing from one of the best merchandise buyer programs in the country, Abraham & Straus, so I have a professional background in many of the disciplines that inventors need help with. Second, I spent many years bringing several highly successful products to market for inventors such as Misto and Smart Spin, which I 100 percent profited only on back-end sales success, which means the inventors paid nothing out of pocket for my help and profited many times over themselves. Third, I studied everything I could get my hands on in the innovation world. I wanted to see who and what else was out there. I read many books and hands-on researched as many inventor company models as I could find. I put myself through a rigorous training program before determining the exact program I wanted to build. Fourth, I decided to focus only on the few industries that I am an expert in and not pretend to be a jack of all trades. And fifth, oh yeah, I almost forgot . . . *I do not charge inventors!*

The Wrong Advice

All of this causes a major dilemma that I witness every day, as I receive thousands of product submissions each year. Inventors are being given the wrong advice, and they are often unprepared to advance their products. The ideas are often half-baked, with many sketching their ideas, or designing on CAD, or making sell sheets without proving product functionality. If I even like an idea, the next question out of my mouth will be, "Do you have a functioning prototype and filed intellectual property?" Somewhere along the way, some inventors have learned to submit this way. They are encouraged to draw up ideas, craft an attractive sell sheet, and try to sell them without really applying themselves. It seems to me, as a seasoned open innovation expert, they are often coached to throw ideas up against the wall to see what sticks, and I can detect this lack of core discipline from a mile away. If you send in several unprepared submissions and waste my time repeatedly, you, your coach, and the collective program will lose significant credibility. Using a hockey metaphor describing the occasional long-range blue-line slap shot that gets through to the goalie, how many result in goals scored? It's the same with inventing. How many ideas without development become successful in the marketplace? The answer is very few.

The Road Less Traveled

It's hard to believe, but a third of my professional life has been spent helping inventors. What I have observed over the years has never changed my views of what so often happens in this industry. It's the reason I set up my unique open innovation model the way I have. Someone recently said I was "lucky" that I didn't have to market to and charge inventors. I think they missed the entire point. My business model is set up specifically to avoid exploiting inventors financially. I studied the marketplace and set up my entire model from scratch to avoid the mistakes that others make. Luck did not enter into my business model. The reason I didn't start taking inventor money upfront was recognizing that honest, sustainable profit comes from actual marketplace sales success. I don't get paid on hope or dreams, nor would I want to. I have been fortunate to earn enough money from the launches of highly successful products and I don't retain employees or an expensive office. I am paid by companies with sufficient resources, so cash flow has never been the driver of my mission.

11

Inside the United States Patent and Trademark Office and Washington, DC, Politics

The United States Patent and Trademark Office (USPTO), located in Alexandria, Virginia, is a federal agency reporting to the US Department of Commerce. It advises and assists the president of the United States, the secretary of commerce, and all other agencies of the US government in matters involving domestic and global intellectual property. If you are ever in the Washington, DC area, you should visit this beautiful facility.

Unbeknownst to many, the original US Patent Office was the first building constructed in Washington, DC, back in the 1790s, right after the White House. The economic importance of patents to our founding fathers and the birth of America cannot be overstated. Thomas Jefferson was the first commissioner of the Patent Office, and George Washington personally licensed the third patent ever issued, which became the basis for the successful grist mill and whiskey business at his Mount Vernon farm.

The director of the USPTO also serves as an undersecretary of commerce. They are selected by the secretary of commerce, who is appointed by the president. So, technically, the United States Patent and Trademark Office reports to the president of the United States, though it follows laws and regulations passed in Congress and examines patents in a manner that reflects the latest judicial rulings, including the Supreme Court.

INVENTOR INDUSTRY ROCKSTAR

USPTO's Elizabeth Dougherty

As the eastern regional outreach director for the US Patent and Trademark Office, Elizabeth Dougherty carries out the strategic direction of the undersecretary of commerce for intellectual property and director of the USPTO, and is responsible for leading the USPTO's East Coast stakeholder engagement. Elizabeth has more than twenty-five years of experience working at the USPTO. She began her career as a patent examiner after graduating from The Catholic University of America with a bachelor's degree in physics and a JD from The Columbus School of Law at The Catholic University of America. She has dedicated much of her career to the USPTO's outreach and education programs focusing on small businesses, startups, and entrepreneurs. With tireless personal effort and passion, she has developed, implemented, and supervised programs that support the independent inventor community, small businesses, entrepreneurs, and the intellectual property interests of colleges and universities. Elizabeth is one of the many good people within the USPTO who cares deeply about inventor issues and education.

A patent is a form of intellectual property that gives its owner the legal right to exclude others from making, using, selling, and importing an invention for a limited period in exchange for publishing and enabling public disclosure of the invention. It's a trade-off between independent inventors and the US government, allowing individuals and companies to profit exclusively from developing unique technology for twenty years, and then turning over the formula to the public. Patents are an exciting merging of personal and public interest that is spelled out in the Constitution. Our founding fathers recognized that incentive was necessary to spur innovators to invest their time and money in ways that an entire nation could benefit from.

Intellectual property also includes trademarks and copyrights, but let's focus on the patent portion, both utility, and design. Patent applications are made up of three main parts:

- Drawings
- Detailed description or specifications
- Claims

Patent Drawings

These are illustrations used to describe, through visual representation, the concepts of an invention. Drawings can be of any physical object or electrical diagrams. Many different types of pictures and figures can be used to explain an invention.

Detailed Description or Specifications

This is the section where the invention is described in a defined manner, detailing the patent drawings at every specific reference point. The purpose is to describe the invention in detail, with enough precision and comprehensiveness so anyone would be able to recreate the invention.

Claims

These are the essential portions of a patent filing. Claims are distinct phrases describing each specific inventive concept of an invention. Claims are the purposeful intellectual property and the only portion of a patent filing that is enforceable. Remember, claims are the actual property used to prevent others from using your patented concepts.

A utility patent may be granted if the subject provides some identifiable benefit, and it is capable of use. It's more technical and valuable in the licensing process than a design patent, which is based upon the unique visual qualities of a manufactured item. A design patent may be granted if the product has a distinct configuration, distinct surface ornamentation, or both.

Patents Today Are Under Siege

With the advancement of today's technology, there's a rift in the innovation community between independent inventors and big tech firms. While our society has embraced technology from the development of high-tech computers and communication equipment to software, the patent subject matter is becoming increasingly more complicated. As technology grows at a rapid rate, the historical foundation of patents and intellectual property has come under scrutiny from the big tech industry. With their tremendous size, they are a significant lobbying force. Their reach goes far beyond what many might realize. New frontiers such as artificial intelligence (AI) are increasing the level of IP complication, not to mention privacy issues such as algorithms attached to social media sites like Facebook and LinkedIn. Our world has changed, and old-fashioned independent inventors with utility and design product inventions are caught in the middle of a bigger corporate battle.

I should note that Big Pharma companies are typically on the same side as inventors and desire a robust patent system, as do old-line companies with large patent portfolios. Many of these companies got their start because of the patent system and grew over the years and remember their roots. Universities tend to be on the inventors' side also. So, it's an interesting division of bedfellows that makes up our current intellectual property tumult.

Leahy-Smith America Invents Act of 2012 (AIA)

The most significant piece of intellectual property legislation to be passed into law by Congress in over a hundred years was the somewhat mistitled Leahy-Smith America Invents Act of 2012. The signature portion of the legislation was the switch of our traditional patent system from first-to-invent to first-to-file, meaning the inventor of record going forward needs to be the one who filed it first, not the one who kept a notebook in their basement, yet never formally advanced an idea. This change put America in sync with all the other patent systems around the world. All in all, there were very few legal cases that went in the inventor's favor based on a notebook drawing, so even though it was a symbolic setback for inventors, that portion has not caused a great deal of actual misfortune. However, another critical ingredient that was cleverly buried within the bill has done a great deal of harm to grassroots inventors.

HONEST TIPS

America Invents Act

The Leahy-Smith America Invents Act (AIA) was passed by Congress and signed into law by President Barack Obama on September 16, 2011. The law represents the most significant legislative change to the US patent system in many years. Named for its lead sponsors, the act switches the US patent system from a "first-to-invent" to a "first-inventor-to-file" system, eliminates interference proceedings, and develops post-grant opposition. Its central provisions went into effect on March 16, 2013.

The Patent Trial and Appeal Board (PTAB)

A more disconcerting part of the AIA turned out to be the creation of the Patent Trial and Appeal Board, otherwise known as the PTAB, which numerically has worked against independent inventors. Every inventor should be aware of this potential post-patent-grant review process. Technically it is an adjudicative body within the US Patent and Trademark Office. The PTAB decides appeals from the original decisions of USPTO patent examiners and adjudicates the patentability of any issued patents challenged by third parties in what is called post-grant proceedings. The concept was to avoid the courts and save everyone's expense when injured parties sought patent infringement cases by having the USPTO take another look at issued patents to see if they ever should have issued them in the first place. Sounds innocent enough. Right?

The problem is the AIA was cleverly written, with heavy tech industry lobbying influence, ensuring any issued patent could potentially be brought back to the USPTO before the PTAB "judges" for a post-grant validity review process. There are today about three hundred judges on call who have no judicial training or technical pedigree. The PTAB judges are selected and approved solely by the undersecretary of commerce and director of the USPTO, who has wide latitude in deciding who is selected and how to direct them. This is one reason why many inventors questioned former USPTO Director Michelle Lee's motives, as she had been a former patent attorney for tech giant Google. There appeared truly little transparency with PTAB hearings and proceedings.

With the PTAB in place, issued patents that have already gone through a rigorous, multi-year, highly detailed examination and approval process, paid for at great expense by the inventor and fairly issued, are now subject to being overturned, often after a developed product has launched commercially. Big money, at this point, is often at stake. Keep in mind, most patents that are challenged in the PTAB process are for products that have had some success in the marketplace, as few products without financial success are worthy of infringement.

Think about it. If you are a big company and you decide to infringe on an inventor's patent and are sued for it, you can now take the issued patent back for review through the PTAB and challenge its validity as if it was never issued. Unfortunately, the same organization that issued your patent, at your expense, can now take it away. This is a massive benefit to infringing companies. There's a term I have heard used to describe this phenomenon; it's called "efficient infringement."

The real killer in PTAB reviews, as opposed to legal courtrooms, is an issued patent would automatically revert to having no validity, just like before it was issued, an underlying assumption provided to all patent holders if they were to enter a legal courtroom. This, despite inventors having gone through years of review and financial expense to obtain what they thought would be a valid patent they could build a company around or license. Additionally, anyone can call into question a patent's validity after it was issued, even if they have not taken part in, or are affected by, the adjudication. On top of all of this, if the PTAB rules in favor of the patent holder, a repeated challenge can be brought back an unlimited number of times before different PTAB judges, unlike a court of law where decisions are final after limited appeals. There are limitless bites at the apple if you are trying to destroy an issued patent. From an independent inventor's standpoint, the process undermines any reasonable financial projection, investment fundraising, and marketplace continuity. At the very least, the PTAB has shaken inventors' confidence in our patent system, and at the most, it's downright irrational—and unfair.

HONEST TIPS

The Patent Trial and Appeal Board

The PTAB is an administrative law body of the United States Patent and Trademark Office (USPTO), which decides issues of patentability. It was formed on September 16, 2012, as one part of the Leahy-Smith America Invents Act. The PTAB Trial Division handles contested cases such as Inter Partes Review, Post Grant Review, Transitional Program for Covered Business Method Patents, and Derivation Proceedings.

Of all the products caught in an infringement dispute, where a request is made of the PTAB for review, about 67 percent are approved for such review. So, only one-third of cases are rejected for insufficient cause. Of those that move forward to PTAB review, however, an unbelievable 88 percent have at least one claim rejected, which invalidates the full force of the issued initial patent, even if the inventor waded through four to five years of USPTO review to be approved in the first place. Remember that only the best products are infringed upon and they are typically the only ones that get adjudicated, but by my count, a whopping 60 percent of all issued patients brought forth are getting invalidated by the PTAB. More than half! This is utterly amazing! Imagine if, after every NFL football game, 60 percent of the final scores were reversed? I think we'd all lose confidence in the process and stop watching the game.

A couple of logical questions come to mind: If the failure rate is that high, what are the original examiners doing wrong? Or, are these rejection numbers so high because of the political clout of the big tech firms? Are the PTAB judges truly qualified, as the entire review process seems to be out of public view? How can anyone raise capital in this uncertain climate for new product launches when the ground can be swept out from under you? Does the PTAB promote or discourage innovation in America? I will let you and others decide that for yourselves, but I wonder what the average American would think if they actually knew what was going on here.

Some experts report a full PTAB hearing can cost up to $500,000, an expense very few independent inventors can afford. The additional challenge is that attorneys are, by and large, unwilling to take PTAB cases on contingency because there is no potential financial reward to be won from an infringer; ergo there is no profit motive for them. The only possible win at PTAB, if one could call it that, is that you get to keep the patent as it was issued initially. Not much of a silver lining. I was in a PTAB lecture at the USPTO recently when the validity issue was explained. It was disheartening, so I had to get up and voice, on behalf of all inventors, my shock and disapproval. Unfortunately, the only real remedy is a legislative fix on Capitol Hill as the USPTO does not make the laws; they only follow them. This will be very difficult to accomplish soon with many other ongoing government priorities.

Capitol Hill and Patent Trolls

One of the great, negative marketing monikers the big-tech firms came up with to advance their anti-strong-patent legislation is "patent troll." Maybe you have heard of it. It's been a marketing boon for them. This term is derisively used to attack nonpracticing attorneys and others who use other people's patents and potentially threaten the big companies with infringement lawsuits. While it is certainly possible that there are bad actors in America chasing ambulances and causing trouble, the real stats show it's a very low percentage of all legitimate patent infringement cases where this might possibly be happening. This percentage has remained roughly the same for over 100 years. Yet, the big-tech companies, and their multi-billionaire owners, would have the world believe through aggressive PR that the problem is rising, and patent trolls will be the death of innovation. They have leveraged this campaign to promote regressive patent legislation. I like to compare their approach to this analogy: If you have some weeds in your lawn, pour gas all over and light it. This will undoubtedly get rid of the weeds, not to mention your lawn also. From an independent inventor standpoint, the big tech, anti-strong-patent legislation has been draconian. Our founding fathers are surely turning over in their graves.

Even More Tech Legislative Attempts

To make matters worse, the AIA wasn't enough for the big-tech companies. Before the ink had dried on the AIA, they were promoting even more legislation onerous to inventors on Capitol Hill. The main advocates were congressional Judiciary members Darrell Issa (R-CA) and Bob Goodlatte (R-VA). They introduced and promoted more one-sided bills demanding extravagant concessions from the patent system, the patent community, and independent inventors. Fortunately, their bills did not advance through the Senate. There are currently new bills recently introduced with the hope of getting the pendulum moving in the other direction. Still, the timing is tough, with so many different front-burner issues. We'll see.

The Courts

The Supreme Court has become involved in patent fights over the past few years. It's clear they don't have much of an appetite for advanced engineering and software concepts, which they do not have a great deal of technical knowledge about. The big-tech firms seem to have reach there also. Many of their recent rulings have gone against the core beliefs of independent patent holders and inventors. In Article 8 of the Constitution, in 1792, "Congress shall have the power to promote the Progress of Science and useful Arts, by securing for limited Times to Authors and Inventors the exclusive Right to their respective Writings and Discoveries." This article provides what should be certain intellectual property rights for inventors. It has been a long-held core tenet and incentive that has promoted inventors to take risks and has fueled the economy of the United States to what it is today. This historical article is now under attack. They are now defining patents as government "franchises" instead of property rights. How could we possibly have gotten here?

The Political Players in the Patent Tumult

For those keeping score, it might make things easier understanding where all of America's innovation shareholders stand in today's patent legislative efforts. Each of these parties has a significant stake in the current patent disputes. Here's where they come down on the vital issues:

1. Big-tech companies are led by Google. They have a great deal of lobbying clout. These companies feel they are being sued for infringement too often, and patents only get in the way of their bringing new products to market. They command substantial market share and no longer require the protection that independent inventors need to get off the ground. These big-tech companies have ridiculous amounts of resources. Some are bigger than countries. Most started with a single patent not long ago, yet they don't seem to remember those days. Now, they look out for what's best for them.
2. Aligned with big tech are the nation's largest banks and financial institutions, along with leading retailers such as Amazon and Macy's. Retailers don't care much about IP; they only want to sell more goods, no matter where they come from, even if the product infringes or is counterfeit. They are all about sales—their own sales.
3. Old-line companies such as GE, 3M, IBM, and Qualcomm, who own large patent portfolios, are generally on the side of independent inventors and strong patent

protection. Most have built their companies the old-fashioned way, through research, development, and filed intellectual property.

4. Big Pharma is the other large wheel in the national innovation picture. Most side more with inventors and want strong patents. They spend a great deal of money on research and development. Additionally, it can take a long time for a product to get to market, so they want it protected for the full twenty years to pay back high research costs before it goes generic.
5. Universities, who do their own research, want to profit from their work and investment and tend to side with inventors. The only major one that doesn't is Stanford, where the school has close connections to Google and Silicon Valley.
6. Inventors are like hobbits living down in the shire. Taking a visual from *The Lord of the Rings,* inventors are little people, well-meaning, growing their crops with these big entities swinging swords well over their heads, but inventors have the responsibility of guarding the ring! Never forget, inventors are the pursuits, the heart of innovation, always disrupting and creating new jobs. Inventors want a strong patent system that protects them.

The Repercussions for US Innovation

Our founding fathers recognized the importance of inventors to America's economy, which depends upon innovation. There are now many reports that China is quickly developing a stronger and bigger patent system than the United States, which has traditionally ranked as the best in the world. Sorry, folks, but regardless of China, our heady days are over. Other countries are now either closing the gap or surging ahead, and it is our own doing. We are shooting ourselves in the foot catering to the special interests of aspiring monopolistic companies that already make a ridiculous amount of profit and are owned by multi-billionaires. Our patent system was designed to stimulate growth, ensure competition, develop entrepreneurship, and fan the innovative American spirit. This is what our founding fathers envisioned. Our country needs to get back to its innovative roots. We are in the process of losing our way.

12

The Non-Profit Inventor Help Mission

The United Inventors Association of America is a federally registered 501(c)(3) non-profit educational organization providing free learning resources to, and advocacy for, the independent inventor community. The UIA (www.uiausa.org) encourages honest and ethical business practices and does not charge inventors, raising money through generous donations from its board and sponsors, as well as hosting special events. As president of the Board of Directors of the UIA, I have worked for many years to promote the non-profit inventor mission.

HONEST TIPS

The UIA Mission Statement

Empowering inventors through education, access, and advocacy: The United Inventors Association of the United States of America (UIA) is a 501(c)(3) non-profit educational organization dedicated to providing resources to the inventing community while encouraging honest and ethical business practices among industry service providers.

Why is a Non-Profit Inventor Organization Important?

As I have referenced in other areas of the book, there are a plethora of companies and individuals who prey on inventors' naivete and false dreams of making millions of dollars taking a product to market when the odds for that type of success are slim. A sea of rough waters surrounds the innovation community with many self-centered sharks who like to turn unrealistic dreams into gold mines for themselves, even if there is no financial benefit for the inventor. Some individuals and schemes look benign enough on the surface, but the numbers and results are dismal when it comes to generating real profit for inventors.

The UIA was initially founded in the early 1990s as an extension of the United States Patent and Trademark Office. The UIA has been around for thirty-five years and was established as a watchdog organization. The primary purpose was to provide a heads-up to the inventor community about predatory companies, agents, and coaches. The idea was to supply free educational information and materials to inventors.

The 501(c)(3) non-profit designation means everything to the UIA. The organization and its board members adhere to rigorous federal charitable guidelines. Some other so-called inventor associations do not, so be sure to ask for details if you get involved with one. Some are merely fronting for profit-driven individuals.

The UIA, with a board of directors and bylaws, provides ethical businesspeople who care about inventors a platform to give back to the community in a charitable way. The UIA Board of fifteen members includes an impressive collection of highly knowledgeable folks from diverse fields, including inventors, patent attorneys, product developers, and open innovation specialists. The UIA holds regularly scheduled meetings throughout the year, with recorded attendance, voting on all significant matters, upholding a strict set of governing by-laws with an executive director and staff. The UIA also adheres to stringent Internal Revenue Service (IRS) regulations. Every board and team member must adopt the 501(c)(3) non-profit mission, put the inventor first, behave appropriately, and hold themselves personally accountable, along with their respectful companies.

Every board member of the UIA is involved in our mission to educate, advocate for, and help inventors get their products to market. Many of our team travel to trade shows and other educational events across the country at their own expense. No one takes a salary or expects to profit from their work. The organization takes great pride in upholding our mission as a 501(c)(3) non-profit. I am proud to serve as the president of the UIA and work closely with some of the industry's and nation's most outstanding leaders. I believe the UIA upholds the innovation standards, spirit, and beliefs of our forefathers. By the way, anyone can join the UIA and membership is free.

Inventor Trade Show Pavilions

One terrific activity where the UIA has found a great deal of success is organizing inventor pavilions at major industry trade shows across the country. Here are a few of those industries that the UIA has been involved with, in one way or another, over the years:

- National Hardware Show (Las Vegas)
- PGA Golf Show (Orlando)

- Made in America Show (Indianapolis)
- International Housewares Show (Chicago)
- Toy Fair (New York City)
- PDMI Direct Response Industry Show (San Diego)
- Toy and Game Fair (Chicago)

Trade shows provide inventors a tremendous opportunity to showcase their new products to interested companies under the right conditions, meet other inventors, gather education from industry experts, and participate in pitch panels. I will cover this more in a later chapter.

Local Inventor Clubs

Throughout the United States, numerous independent local inventor clubs meet one evening each month. These clubs are the heart and soul of innovation in America. I am a firm believer that every inventor should join their local club and network with like-minded people. There is so much to learn from others who share your journey. I describe club meetings as real boots on the ground. Most of these clubs are easy to find online, have terrific leaders, and offer regular speakers. On the UIA website (www.uiausa.org), we list many of the local inventor clubs around the country, their contact information, and locations. Our core UIA mission includes working closely with and supporting all the local inventor clubs throughout the country. You will know a local club is a good one if it is listed on the UIA website.

Pro Bono Patent Assistance Program

The Pro Bono Advisory Council, a 501(c)(3) non-profit board that I also sit on, is the group that assists the USPTO administering our nation's pro bono patent assistance program. Inventors and small businesses that meet certain financial thresholds (and other criteria) may be eligible for free legal help in preparing and filing a patent application. The Patent Pro Bono Program is a nationwide network of independently operated regional programs that match volunteer patent professionals with financially under-resourced inventors and small businesses for the purpose of securing patent protection. Each regional program provides services for residents of one or more states. Your gross household income should be less than three times the federal poverty level guidelines, though some regional programs may have different criteria. You must also demonstrate knowledge of the patent system by having a provisional application on file with the USPTO, or successfully completing the certification training course and describing the particular features of your invention and how it works.

INVENTOR INDUSTRY ROCKSTAR

Pro Bono Pioneer Jim Patterson

Jim is a founding partner of the law firm Patterson Thuente IP, a full-service intellectual property boutique based in Minneapolis, whose practice encompasses all aspects of intellectual property law. Jim's clients are leading companies and individual inventors, located around the globe and in a variety of industries, including medical devices, defense, electronics, plastics, and mechanical equipment. Jim is the main person behind the recent creation and launch of the 501(c)(3) non-profit Patent Pro Bono Advisory Council, which he chairs. The PBAC works closely with the USPTO to oversee our country's national intellectual property pro bono program. The result provides patent access to low-income inventors. Jim's time and charitable work benefiting the inventor community has been invaluable.

USPTO Invention-Con and UIA Club Leader Conference

Each year the USPTO hosts an immensely popular educational gathering for inventors in Alexandria, Virginia. The event lasts two days and features USPTO and outside experts and speakers. The UIA also coordinates a national educational forum around this event at the USPTO specifically for inventor club leaders, featuring two days of educational speakers, networking, and then attending the USPTO's Invention-Con event. In 2019 we had nearly fifty inventor industry leaders participate from all over America, who spent time with one another learning new ways to outreach, take advantage of SBA government programs, and hear the latest about intellectual property news. The UIA serves as the primary inventor community liaison to the USPTO.

Capitol Hill

We also spent time on Capitol Hill with the club leaders visiting the offices of many Senate and House members, advocating for why inventors are crucial to the country's economy, and how they support grassroots innovation. As a national organization, we feel it's vital that every local club leader also understands the national challenges inventors face, and we want to help them to serve their members better.

INVENTOR INDUSTRY ROCKSTAR

Minnesota Inventors Congress Deb Hess

Deb is the former Program Director of the legendary Minnesota Inventors Congress (MIC) and today uses her marketing and management experience to educate inventors about developing marketable products. She is known for her passion and for helping aspiring inventors learn how to make important business decisions launching their new ventures. Deb has been supporting independent inventors for over thirty years. She is a certified trainer, coach, and facilitator in fostering the creative and innovative spirit in entrepreneurs and is invited regularly to speak to inventors, students, intellectual property law students, entrepreneurs, manufacturers, and others. On an international level, she has served as a juror for the World Cup of Computer-Implemented Inventions in Taipei, Taiwan, and as the US Representative for the Global Sustainability Conference in Bangkok, Thailand. Deb currently serves as the treasurer and operations director of the United Inventors Association and is an indispensable part of the UIA non-profit team.

Other Inventor Organizations

There are a few other national inventor organizations, two of which are non-profits that the UIA supports and works with, and the third is a cleverly disguised for-profit:

US Inventor

US Inventor is a 501(c)(4) non-profit, allowing them to raise funds for lobbying. US Inventor is very politically active on issues such as patent reform and national intellectual property rights and legislation. Led by tireless workers and passionate inventors Paul Morinville and Randy Landreneau, US Inventor recently introduced a bill on Capitol Hill to protect independent inventors. The UIA supports their efforts, and they serve as a nice complement to the UIA because they lean more toward political activism, while the UIA is more involved with grassroots education. Membership in US Inventor is free.

Inventor's Project

The primary mission of Inventor's Project is to promote innovation and defend intellectual property. Co-director Charlie Sauer is an economist and policy specialist who has spent lots of time on Capitol Hill, and worked for a governor and an academic think tank. He is also an entrepreneur, president of the 501(c)(3) non-profit Market Institute, and author of *The Profit Motive*. Charlie has been featured on Entrepreneur.com, Fox Business, the Daily Caller, and Women Entrepreneur. In addition, he has also written congressional testimony and many speeches for politicians, business owners, and academic leaders.

Inventors Groups of America (IGA)

A self-professed inventor organization is called the Inventors Groups of America, although it is also referred to as America's Inventors Groups. The IGA is an extension of the for-profit inventor coaching company inventRight, generating new inventor leads on behalf of their for-profit coaching services. The IGA counts upon the goodwill of local non-profit inventor clubs around the country to generate cash flow for inventRight. Because they are not a 501(c)(3) non-profit, they are not required to adhere to rigorous federal guidelines and IRS standards; nor do they have an elected board of directors with scheduled meetings and minutes; nor have they developed transparent policies utilizing bylaws and ethics guidelines. The IGA is a lead generating pay-to-play platform for a coaching business model.

13

SIAM

This chapter is highly personal to me. For some inventors, the following may be the most crucial chapter in the book. The reason is if you invent enough you are bound to fail. How you deal with that may make all the difference in eventually succeeding. I want to publicly share what I have learned over the years to cope efficiently, as a highly energetic risk-taker, with the flip side of success, which is frustration, anxiety, and even unabashed failure. The inventing world is riddled with failure, so somewhere along the way, you will most likely experience some sort of pain that should be addressed.

It is difficult to accomplish great things without taking big risks. Some folks avoid risk at all costs. Some mitigate risk through planning. Then again, some are utterly reckless. Whichever you are, you will inevitably fail along the road of life—or at some point, carry an untenable burden. I don't think anyone is immune to the heartache and challenges of life. How you deal with failure and stress will both define who you are as a person, as well as return you to the saddle and get you back on track with your mission, sooner rather than later. This is especially true of risk-taking inventors.

High-energy, anxiety-filled, attention-challenged people in their youth, like me, have always had a hard time growing up while fitting within society's norms. Many times, institutionally, we are the square pegs jammed into round holes, the ones who had a hard time sitting still in classrooms and had a hard time focusing. We are the ones who say what's really on our minds, sometimes awkwardly at the wrong time. We are the ones whose teachers didn't fully understand us, so we were continually punished or put down. We are the ones directly descended from the hunters in prehistoric times. We are the disrupters, thinking differently, not quite content with business as usual. We learn far more from doing than reading how to do it in books. We're usually the pioneers. If you are a real inventor, does this sound familiar?

For everyone, growing up, we learn how to cope with and adapt to the world around us. For those socially gifted, this may be natural and easy. For others, it's not so easy. Let's face it, living in Western civilization can generally be stressful. The speed of the game, high technology, eight-lane highways, skyscrapers, and a million things to think about every day; it can be overwhelming. The excellent news for crazy disruptors is, if we work hard to figure everything out and somehow adapt, we generally bring to the table a whole new dimension of thinking, sometimes way out of the box. I like to think that the energy that gets us into trouble as kids is the very same lifelong energy that, once harnessed, can lead us to unbelievable accomplishments and innovation breakthroughs. The challenge, quite frankly, is staying alive and corralling our unique inner current. It turns out that if we can get to adulthood, there is so much we have to offer.

Many years ago, I developed my own internal code of SIAM to address anxiety and better deal with life when struggling in business or life challenges. I have lived by this system ever since and would like to share it with you. Through a tremendous amount of hard work, I was able to quantify the conditions that generate fear and anxiety and determine the underlying factors that needed to be addressed. I eventually discovered that many streams feed into a major river, and to handle the more significant challenges, I needed to understand everything about the river. The long-term benefit is that I became truly aware of who I am as a person, what my strengths and weaknesses are, and how to help others when they similarly struggle.

SIAM is an acronym for "S" TOP . . . "I" DENTIFY . . . "A" DDRESS . . . "M" OVE ON.

Stop, Identify, Address, Move On

I have learned that when my anxiety bell goes off, it usually means something is not quite right, and things need to be recalibrated. When you know the signs, and can harness the emotions, it acts as a wonderful early warning system. When anxiety hits, I immediately *stop* what I am doing (or as quickly as I can without drawing attention) and take a break. Sometimes I close my eyes and look up into the sky and breathe deeply. I think of a relaxing thought. I manually shut myself down. I have become quite adept at this, and so can you with practice.

■ ■ ■

HONEST TIPS

Definition of Anxiety

Here's the simplest and best definition for anxiety that I have ever heard: "Fear of the unknown, multiplied by how important it is to you."

The next step, when I am calm, is to mentally search and rationally *identify* the real issue behind the anxiety. All of this can be challenging because I've learned the real causes are not the ones that always seem apparent on the surface. Usually, it's not the obvious issues like financial worries, or an angry boss, or getting on a plane, or even giving a speech. Often the real issues are subliminal. Maybe your wife did something to trigger a mom flashback to youth, or perhaps you feel you are simply not living up to your expectations because you had a tough dad who drilled you growing up. Maybe you did something unethical that you know deep down is not right and are struggling to come to grips with it. Whatever it is, with lots of practice, this identification step will become easier over time. As you will begin to understand, a light bulb will go off, a breakthrough if you will, and things will start to click into place internally.

The next essential step, when ready, is *addressing* the problem. Once you know what's bothering you, create a way to address it proactively. I'll provide a deeply personal example. When I grew up, I was the third of four boys, and extremely hard on my younger brother at times. Later in life, I felt tremendously guilty about this, and as a result, I began taking responsibility internally for everything he, in turn, struggled with. I believed subliminally if I had just treated him right in our younger days, he'd be better off. All this is an immense weight for anyone to carry around. So, one day, while in my twenties, I asked him to join me for lunch at a New York City bistro. Once there, I apologized profusely for my childhood behavior, and with that came a flood of tears. My brother, of course, was embarrassed with all the emotional attention, particularly with others seated around us. He had never seen me like this before and assured me that I had been a great older brother, and to stop acting like a baby. The next day, I never felt better in my life! A cloud was lifted, which left me at peace. It was a wonderful reconciliation.

The last critical step is to *move on*. Once you have identified and addressed your challenge, it's time to move on. Don't straddle the problem or look back. Just *let it go!* The result of this continuous reconciliation process is that instead of anxiety being a curse in my life, it today serves in many ways as an early warning system that lets me know subliminally when all is not right, and things need correcting. It's an alert telling me I need to stop, address my problems before they get away from me or add to other pressures, which can become disabling. SIAM is like a self-cleaning oven. I no longer go to sleep carrying problems without identifying them and developing a game plan to address so that I can feel at peace. By doing these enough times, one day, I arrived internally at a life filled with peace.

SIAM is not just an acronym for me; it has become a personal code that I live by, every day, and a way to understand myself better. Over the years, I learned that I'm a type A personality and an overachiever. I have confronted challenges head-on, never backing down, much like harpooning a whale and going for a Nantucket Sleighride. Mostly the rewards were great, but sometimes the failures were epic. Knowing who you are and your limitations can go a long way toward relating well with others and eliciting the most from your abilities. I learned this the hard way.

How SIAM Can Help Inventors

So how can SIAM specifically help inventors? For starters on a personal front, I set out to write this book for the joy and peace of giving back in a meaningful way to others. It's part of my own "address" process. When I look at the inventor community, I see some large companies who want absolutely no restrictions on taking the intellectual property of others. I see a government and nation that does not fully understand the value of individual brilliance, disruption, and independent innovation. Nor do they fully understand the value proposition of rewarding inventors for their risk and sacrifice, which helps American society. They seem to think that innovation comes out of thin air, believing in 20/20 hindsight that every invention was inevitable. They even did this to the Wright brothers for God's sake—the patent naysayers back then proclaiming after they took flight that anyone could have figured it out. I witness invention marketing companies providing mediocre help and the latest generation of America's hucksters more consumed with their own financial success cleverly charging others without inventor profit. Every day, I see many independents trying to get started in this minefield environment.

In my own way with this book, I am trying to stop, identify the real challenges that inventors face today, introduce and address solutions, before moving on. Hopefully, in this way, I will have given back to others as much as has been given to me over the years. This book may not necessarily generate a lot of royalties or adjunct for-profit coaching inventor-help programs, things I really don't care about. Still, it will provide me with peace, and hopefully identify the real challenges facing inventors. So, to the extent that SIAM helped create this book and this book helps you, it's of tremendous value to me.

More importantly, how can SIAM help you individually on your path to innovation? Well, here's a deep secret that other authors who charge for their services may not have a personal financial interest in fully explaining. If you are serious about inventing—you will fail! Inventors, by definition, are risk-takers, and many ventures do not work out. Tom Risch, the inventor of Misto, told me once he had two winning products among sixteen failures, some colossal. Josh Malone, the inventor of Bunch O Balloons fame, failed multiple times with other products before hitting it big. And these are two extraordinarily successful inventors! What about all the others you've never heard of? While middleman marketers may tout that inventor gold mines are miraculously around the corner and pony up your cash, understand in the real-world most of your invention efforts will fail. Being a professional inventor, how you handle failure will define you. How you get back up off the mat, learn from your mistakes, and carry on is who you really are.

I have partnered or been involved with the launch of many phenomenally successful products, operate multiple open innovation programs, started nine businesses, and earned a substantial amount of money over the years. However, I have not made it through life without serious challenges, which led me at times to the brink of financial despair. I had one business, even after I sold and liquidated everything, that I still owed nearly a half-million dollars. I worked that debt off over several years, which seemed at the time like it took forever. I did stand up and address the challenge, though, and I learned a great deal from it. And this is one of the principles of SIAM, which is being accountable in ways that bring eventual peace.

My biggest invention disaster was Stir Chef, the Saucepan Stirrer. The device was clever because it folded into a cylinder about the size of a small coffee cup, which made it convenient to store in a kitchen drawer. It had a strong motor that was powered by four AA batteries, and featured fold-up spring-loaded clamps that adjusted to fit over the sides of any size saucepan. Once attached, it contained a drop-down paddle that stirred anything in the pot. I felt stirring is one of the most used commands in cooking, and this product was bound to be a big hit. I introduced it at the 2005 International Housewares Show in Chicago, and it was quickly acclaimed as one of the greatest new product launches of the year. We were featured on the *Today Show* and took orders for over 100,000 pieces at the show from all the major retailers. I went on QVC myself many times at all times of the day and night to demo it live.

The problems began when we ramped up production and manufactured over 250,000 units in anticipation of sales, reorders, and success. Unfortunately, those anticipated reorders never came, and the product did not sell. It bombed, and when a product does not sell, the retailers shut down and do not pay for any of their invoices. We had no alternative products to exchange, and so the legal issues with these big players began. Next, we were forced to liquidate 150,000 units, which had little value. I spent nine months, every day for nine hours, in a windowless warehouse selling the unwanted inventory for well below cost. Trust me, it was a humbling experience. I left a lot of blood, sweat, and tears on that warehouse floor.

HONEST TIPS

Stir Chef, a Hard Lesson

In 2005, I spent a year with partners developing an automatic saucepan stirrer that was absolutely perfect for making risotto, as well as helping senior citizens and disabled folks stir. Unfortunately, it otherwise had a very limited market. Buoyed by a great introduction at the International Housewares Show, we manufactured 250,000 units and sold the first 110,000 rather painlessly. We even knocked off ourselves with a similar version called EZ Stir for discount retailers. Neither version sold through at retail. The mistake: Both were well-engineered products, but never properly consumer tested.

It turns out that high-end chefs like to put their own love into what they are making and stir themselves, and low-end novice cooks don't really stir. Two years of my life on the surface were wasted. I lost a great deal of money and generated a lot of stress at home. The only silver lining: The tough lessons that I learned were invaluable preparation for everything I have been driven to do since.

Here's where SIAM came in. First, it was essential to stop, take stock and reflect on what went wrong, and learn from the experience. Many times, inventors are headstrong, filled with dreams, and barrel ahead no matter what the danger. For me, it was crucial to identify real issues and develop a rational business plan. In the case of Stir Chef, that plan ended up being liquidation, a severe yet necessary measure that would entail identifying potential buyers and moving out the stock. This plan did not happen overnight, yet I followed it, one step at a time, like a trail of bread crumbs out of the woods.

The lesson for inventors is that even if you think you have a hit product, you never actually know until it's selling, and real consumer demand generates reorders. The same goes for licensing deals. You may sign an agreement, but until the company does a great job marketing and selling, you will have absolutely nothing to show for it. Licensing agreements are like patent certificates, great to put up on the wall, but unless they are generating positive cash flow, they are mostly ornamental. So, when you if you get into trouble, from time to time, *stop*, *identify*, *address* before *moving on*. Not everything that you work on will be successful. Promise yourself to stop and think about your family, your bank account, and your real prospects.

Failure is tough to overcome, especially when you want to remain passionate. I understand the need to feel that little extra ounce of creativity to be special. I get it. Just be rational. Use the SIAM process and take the necessary time to reflect upon whether you are on the right track. You can accomplish this by listening to real facts and identifiable demand, not phony marketing pitches.

SECTION THREE

Inventors "Be . . . Aware"

14

Industry Trade Shows

There is nothing more educational for the serious inventor than attending an industry trade show. I am a firm believer in the power of networking and getting in front of real people and companies. There is a lot to learn from walking the aisles, seeing all the new products, asking questions, mingling with company professionals, and hearing from industry experts. When you are ready, target and attend a trade show in the industry that you hope to exploit. Hopefully, the time of year is appropriate for your schedule and the location accessible.

I'm not sure if everyone in the invention industry truly understands the value of attending trade shows. Most didn't come up through the buying ranks, nor do they have significant experience walking shows as a professional buyer. I recently read a narrow-minded business article that argued trade shows are not a safe place for an inventor to show their product, for fear of it being knocked off or stolen. I'd say this advice is very shortsighted. With the proper planning, there are plenty of ways to protect oneself—but then I realized the author had no real experience in this arena. At the very least, I can promise you will be a whole lot smarter after attending a trade show, and you might also walk out with the framework for a licensing agreement. I have seen that happen many times. In fact, I have arranged licensing agreements for products that I discovered at shows.

Trade shows are ground zero for most industries and represent in one location, and one snapshot in time, everything new in an entire field. They represent an excellent opportunity to collect, in one place, information from almost every company around the country, or sometimes the world. Most big trade shows are held once a year at convention halls located in major cities throughout the United States, though some industries hold more than one, or even host smaller satellite shows in regional settings. There are also many overseas tradeshows.

Firsthand Experience and Research

I have talked in past chapters about the importance of vetting the market you are inventing products for. This research should include exploration of online stores and walking retail store aisles while examining products and their packaging. Trade shows, however, are one of the most cost-effective methods for locating all the companies within an industry looking for new products. Trade shows are an integral way to build relationships and get your product in front of the right people, an experience you cannot obtain from a computer screen or email. Trade shows represent an old-school approach to marketing yourself and your product while learning about your competition and gathering new development ideas.

Showing Your Prototype

If you have a booth at a tradeshow, a prototype is a constructive way to showcase your product. Make certain you have patent protection, as you will be publicly disclosing, and keep an eye on the prototype. Outsiders are not allowed to take photos, so be sure to enforce that rule. If you are simply walking a trade show, you will need to decide whether you should bring along your prototype. I might suggest you bring a video instead and then only show it if you feel comfortable. Do not leave a prototype with anyone without signing an NDA. It might be hard to lug around a sample all day, and often industry folks are busy and hectic. If your prototype is not polished, it might not present well in that type of setting. At shows, you are mostly trying to make introductions, not finalizing the sale, so unless you have taken a booth in the inventor pavilion and are aggressively working on initiating a licensing agreement during the show, keep things close to your vest. And always remember that meeting people at the show is as much about presenting yourself well as your product.

HONEST TIPS

The National Hardware Show UIA Inventor Spotlight Pavilion

This special trade show area at the National Hardware Show every May in Las Vegas, sponsored each year by the United Inventors Association, features products not yet on the market, providing an inside look at what may be the next big product or service to revolutionize the home improvement industry. The Spotlight Pavilion, with approximately 175 participants, is now the biggest inventor booth show in the country. Inventors who showcase new products there have the unique opportunity to meet with buyers and potential investors to get feedback on their products, while participating in educational pitch panels, listening to expert speakers, and receiving counsel from national inventor industry leaders. Even if you don't have a booth, it's worth a visit to walk and observe.

Navigating a Trade Show

You can find specific trade shows in every industry by searching the internet, locating the right website, and reviewing the attendee directory, which will tell you all the companies that are exhibiting. They will typically supply a floor plan, so you may plan which companies you want to visit. Use your time wisely because many shows are large, sometimes taking up multiple buildings or halls. You can walk the aisles in order, but often shows are organized by smaller categories, and you can focus on those areas as well. One year I walked over nine miles in a day at the SEMA auto show, so bring a pair of good walking shoes.

If money is tight, sometimes you can walk a show in one full day. It may prove to be a long day, but I promise you will emerge energized. At a trade show, always be professional, which means being polite and patient. Understand most of the personnel in the company booths are working and sometimes have other essential appointments. Those folks you see congregating may be buyers from Walmart or another major retailer, so be considerate. Work the show to investigate what new products companies will soon be bringing to market. This might help you make additional product development decisions. Learn about the latest industry trends. The information you glean could help you narrow in on the best function of your product.

Some of the biggest trade shows in America, and overseas, are massive, like the Consumer Electronics Show and SEMA auto show, both held in Las Vegas. Sometimes the smaller industry shows are broken up regionally, such as the Fancy Food Show. Most shows are not open to the public, so you will need to register in advance as a buyer or trade show guest, which may cost up to a few hundred dollars. Trade shows are mostly designed for retail buyers to walk, see all the new product offerings, and place purchase orders. These buyers are representing what consumers desire, so be sure to approach the shows mentally like a buyer, particularly when looking at the new product areas. Sometimes smaller shows are open to the public, but this is generally the exception. Make sure you check in advance.

Among professionals attending shows, you will also find a mixture of industry salespeople, press personnel working for trade magazines, TV correspondents, international attendees, as well as other inventors. Trade shows include a plethora of interesting people whom you can meet and ask questions in a variety of ways. They are wonderful, spirit-filled events where so many people share a common interest. There are also educational seminars and events worth attending. You will cover more ground in a few hours of walking a show and listening to industry experts speak than you will be sitting at home looking at your computer for a month.

Inventor Pavilions at Trade Shows

More and more industry trade shows are featuring inventor pavilions. In these pavilions, there are booths for inventors who have developed prototypes, filed IP, and are ready to share their inventions with industry experts and companies. It's a fantastic way to engage with your industry of choice. You will meet more connected people in a day than you will in a year at home. One of the best such inventor pavilions is hosted at the National Hardware Show in Las Vegas each May, where the United Inventors Association sponsors over 175 booths and three days of continuous pitch panels. Industry expert speakers provide inventors with an unbelievable opportunity to learn about licensing or taking their product to market directly. Virtually every manufacturing company at the show sends its teams to walk the inventor pavilions looking for new products to license. The Consumer Electronics Show, the International Housewares Show, the PGA Golf Show, and many more now host such pavilions. If you are interested in participating, you should seriously investigate how to engage. It's a tremendous learning experience.

INVENTOR INDUSTRY ROCKSTAR

UIA Inventor Trade Show Pavilions Cathie Kirik

Cathie has spent nearly forty years supporting intellectual property protection and independent inventors; ten years with the Library of Congress Copyright Office and nearly thirty years with the US Patent and Trademark Office. While working at the USPTO, Cathie was an integral part of creating the Office of Independent Inventor Programs, whose focus is to provide educational resources to the inventor community. Highlights included the USPTO Inventors Resource Page, online chats, newsletters, and an established complaint process for inventors working with invention promotion scam companies. Cathie also coordinated the USPTO's Annual Independent Inventor Conferences, which today is called Invention-Con. Upon retiring from the USPTO, Cathie joined the United Inventors Association to continue her passion for supporting the inventor community. She currently serves as trade show director. Her experience and generosity to the UIA are critical to the success of the inventor trade show program.

Meet Licensing Professionals at Trade Shows

At trade shows, you can step into the real world and learn firsthand about your marketplace. I have been attending shows for forty years and still enjoy visiting some of the largest trade shows around the world. The best way to find other potential licensing opportunities is to ask people at manufacturing booths whom they use or might recommend. This is something you can best do in person, on the ground through networking. You can meet company management and sometimes even the presidents. Researching companies is not an easy task, but at trade shows, you can learn firsthand which companies have open innovation programs. You can quickly find out who the major players are, as well as those who shut their doors to outside innovation.

Other Trade Resources

Every industry trade show I have ever attended has a wealth of educational materials, conferences, awards, speakers, panels, and new product platforms. Be sure to schedule these events into your schedule. Typically, after industry experts speak, there is an opportunity to meet and talk afterward, or at least exchange cards and numbers. Keep notes as you will want to reconnect with the people you meet later. Also, take a copy of industry trade magazines and get subscriptions for the year, then read them from cover to cover.

You'd be surprised at what you can learn from the classified ads in the back.

Being Professional

Visiting different booths of the companies that you target is an excellent way to meet people within the industry you are trying to learn. Remember to be polite, introduce yourself, ask questions, and when the time is right, provide them a summary of your product. Don't disclose everything at once. Be cordial and ask how you submit more information about your product after the show. Get their business cards and try to establish rapport. However, don't linger or annoy anybody because most people are doing business at these events. If you come away with the right contact information, you have taken the first step.

INVENTOR INDUSTRY ROCKSTAR

New York City First Responder Chris Landano

New York City firefighter and first responder Chris is both a prolific inventor and founder of the Queens, New York Inventor Club. Recently, during the coronavirus outbreak, Chris started InventorRescue.com with the intention of helping inventors to learn more and help them avoid the costly mistakes he made. The core mission is to help coach and mentor first responders (nurses, EMTs, paramedics, cops) who have a product idea, or are in the process of developing an idea, at no cost to them. Chris wants to connect them with honest and vetted service providers, attorneys, engineers, licensing experts, and manufacturers without charge. Chris is sincere about doing his part to keep inventors from being financially taken advantage of through education and identifying the key warning signs. Chris has also given charitably of his time to the United Inventors Association over the years, particularly helping with the setup and oversight of the National Hardware Show Inventors Spotlight Pavilion.

No Shortage of Trade Shows

There are over thirty thousand trade shows across America each year, and you can surely find at least one in your product category. Research where they are located and plan to attend. Going to a trade show must be well-planned in advance, so you need to determine the best show and arrange accommodations along with financing your trip. Once there, map the show floor and discover where every company booth is located, because most of these shows can be intimidating the first time. A map and a schedule will save you time and energy.

I would strongly suggest instead of spending thousands of dollars on help or coaching from a middleman marketer, to invest the same money, or less, in attending a real-world trade show that will provide you with invaluable business insight. It just takes some curiosity and get-up-and-go.

15

Crowdfunding

For serious inventors, we have discussed the need to go well beyond throwing an idea against the wall to see what sticks. Real product development, pursuing a licensing agreement, and going to market directly on your own, requires varying degrees of capital and a lot of hard work. Everything, including building prototypes, securing materials, filing patents, developing marketing materials and demonstration videos, requires financial investment. Even if your goal is simply to license, you still need to get your product to a point where it will be taken seriously, which will provide you with the best odds for success. Sometimes the capital you need to get started is difficult to secure without giving your company away, hitting up family, or dipping into the kid's college fund. This is where the relatively new concept of crowdfunding might be a great alternative to consider.

Sites such as Kickstarter and Indiegogo offer an exciting new approach to innovators trying to raise capital and get their product off the ground while reducing their own financial risk. Crowdfunding is a way for inventors to keep the assets of their invention to themselves by avoiding angel investors or banks who require collateral. It is a platform where your product is offered for sale in advance of manufacturing and production. These platforms allow you to take pre-orders until sufficient capital is raised and then ship your new product to crowdfunding backers in advance of the rest of the world. It's a quid pro quo: You receive start-up capital, and your early backers receive a cutting-edge product before the prices go up with eventual marketplace demand.

Crowdfunding, as we know it today, was implemented successfully in 1997 by a British rock band. The band told their fans they needed help funding the reunion tour, and they raised money through online donations, which made the tour possible. A company called ArtistShare was the first official crowdfunding platform, begun in the year 2000. The platform was a smash hit, and other notable crowdfunding platforms emerged over the years, starting the explosion within the industry every year since. Today, the crowdfunding industry is an attractive option for inventors because, with crowdfunding, inventors can validate their ideas and increase exposure, while securing seed funding.

Raising Capital

As an inventor, where do you go to raise capital? There are traditional ways to raise funds, starting with borrowing the money from a bank. The only way that will work is if you put enough of your assets up to collateralize the loan. That will likely mean using your home equity unless you have cash or CDs to pledge. Banks are not going to take on any risk of an invention they know nothing about and displays no verifiable means of income. You could tap into your retirement money if you are beyond a certain age; otherwise, there is a tremendous penalty. You could use high-interest credit cards, but you are going out on a limb and opening yourself to enormous financial risk in an arena that is highly speculative.

You can also speak with angel investors who will assess your product and help by lending you the funds to get started. Funds are provided, typically in return for an equity position in the enterprise, which means inviting someone else to take a part of any financial gain from your work. If you end up licensing, this could cut into your potential royalties. They can save you from economic pain on the downside, and perhaps offer expertise in specific areas to get you started, but if you are successful, they get back a lot in return.

You might also go to your family, though some may not have the opportunity. Quite frankly, family members can be as strict when it comes to lending money as angel investors. It can also ruin a close relationship if things go south.

I know there are various books on funding and finding hidden money. An excellent book by Kedma Ough on raising funds is titled *Target Funding*, providing real-world insight on how to find capital for startup businesses. There may be a multitude of ways to acquire funds for your project, but it is up to you to find the way that works best for you. Whatever method you choose, make sure you plan your future moves before proceeding.

INVENTOR INDUSTRY ROCKSTAR

Crowdfunding Guru Roy Morejon

Roy is an entrepreneur, digital marketer, and crowdfunding expert. He started as a consultant for AOL, Microsoft, and DELL Computers before the age of nineteen, and now brings over twenty years of experience helping companies and individuals increase their sales and exposure. Roy is the person behind well over one hundred million-dollar-plus crowdfunding campaign raises, more than anyone else in the field, including Josh Malone's Bunch O Balloons. It's a brave new world launching products these days, and I thank trendsetter Roy for helping me get up to speed on many of the dynamics.

Crowdfunding Beginnings

Crowdfunding is one of the most exciting new platforms to enter the innovation space in a long time. It has been a helpful tool for many worldwide both to raise capital without personal risk and to learn if there is any consumer interest in your product. When it works well, crowdfunding is an excellent method for innovators to exploit their products while selling their wares to the world. The platforms are a perfect way for innovative people to be immersed in a culture of innovation in America. They also provide a way to engage with a community of cutting-edge consumers ready for new products.

The rise in crowdfunding has been rapid. While generating $1.5 billion in 2011, it is expected to skyrocket in growth to over $300 billion by 2025. Since crowdfunding appeared on the map, there are now over 191 different crowdfunding platforms in operation. Some of the largest are Kickstarter, Indiegogo, and GoFundMe. All the significant platforms have seen substantial growth.

On the flip side, preparing for a crowdfunding campaign has become quite sophisticated, requiring substantial investment to shoot a professional demo that will air upon launch. You will need to be far enough along development-wise to convince knowledgeable consumers that you are onto something unique, and they should back you. On some platforms, if you do not raise your target goal, you will not receive any funding and, as such, will be out of pocket with nothing to show for it. You will also need to coordinate a social media campaign to drive sales, without which your product may never sell and raise money. There are companies today that can help guide you through the many challenges, but they need to be rewarded for their efforts. Remember, the crowdfunding platforms also take a percentage of the dollars you raise.

There is one additional challenge to crowdfunding. Crowdfunding sites are online platforms where people from around the world monitor new products and trends. If your product performs well, you can be certain notice will be taken by the competition. Once this happens, you will be on the clock to get to market quickly. During this time, reputable companies may call you to inquire about licensing—and there are also nightmare stories of knockoffs. Just know if you raise hundreds of thousands, or even millions of dollars, the action, on all fronts, will escalate big time, so be sure to prepare for potential success.

HONEST TIPS

Coolest Cooler

The Coolest Cooler was a multi-function, transportable cooler that was initially funded through the crowdfunding website Kickstarter. In the summer of 2014, Ryan Grepper, an inventor, and well-meaning guy, raised over $13 million, making it the most-funded Kickstarter campaign in history at the time. Crowdfunders were offered the product at a discounted rate, but there were problems with orders being fulfilled from the beginning. Because of the low price, it cost more to build the product than what was taken in. In 2019, the company was closed, having never delivered coolers to an estimated 20,000 of the original Kickstarter backers. The lesson for everyone: If you proceed with a crowdfunding campaign, you need to project your manufacturing costs and be ready for success.

Each Crowdfunding Platform's Model

Visit each crowdfunding platform's website and familiarize yourself with their details. Browse the product selections, trade categories, price points, and study the many creative ways in which they are presented. They are all unique, with different rules. Here are the major players:

Kickstarter

You have complete creative control over your project, which will allow your personality to shine. With Kickstarter (www.kickstarter.com), each project has a financial fundraising goal that you set and a deadline for the funds to be raised by. If the goal is reached in time, the individual backers who pledge to purchase are charged the amount they committed to. Kickstarter then takes its 5 percent fee off the top and turns the rest of the money over to you. However, if the project doesn't hit its fundraising target goal, no backers are charged, and the project either dies, or you can regroup, learn from any mistakes, and try again. With Kickstarter, receiving funding is an all-or-nothing scenario.

Indiegogo

This platform helps inventors by providing ample support for each project through information, guidance, and the promotion of successfully funded products. Indiegogo does not include a funding goal. If the projects meet the deadline, a service fee of 5 percent is charged for the money raised. There is also a small financial processing fee.

GoFundMe

This platform is growing in strength and has raised over $5 billion for people worldwide. GoFundMe is primarily used to raise money for community and social awareness issues as opposed to consumer products. They do not set a target funding goal, but they do charge a 2.9 percent fee to process the funds raised after the campaign.

Crowdfunding Readiness

If you are going to pursue crowdfunding, you need to be prepared before you launch, meaning you need a functioning prototype and professional video presentation. This requires investment, but is well worth it as your visual presentation will be critical to backer acceptance and early sales. A poor video can ruin your entire launch. Be sure to go on the sites and review other products to understand the parameters for success better. You will need to budget for an impactful social media support campaign that gets the word out virally and builds sales. It's important to get off to a fast sales start, as campaigns generally only run about a month.

You also need a plan for where you are going to manufacture and what the costs will be. This will determine how much you need to set as a fundraising goal. Typically, you will announce during your launch when the product will be available for shipping and when investors can expect to receive it. If you are not upfront with the consumers that back you, it will create a lot of pressure down the road fulfilling those early orders. Being ready for prime time is essential because once you launch your product will be broadcast around the world.

Crowdfunding Advantages

In my opinion, crowdfunding is a brilliant model, and I don't say that about too many inventor industry platforms. To promote your invention, raise money to manufacture it, and eventually take it to market is a phenomenal opportunity. With crowdfunding, you can acquire financing without collateralizing your home or borrowing. I have seen, on numerous occasions, inventors spending their life savings trying to get a product to market—and leaving them financially discouraged after failing.

Crowdfunding is also the perfect platform to expose your product to targeted communities looking for products like yours. The world is a stage, and by placing your product in a specific category, crowdfunding directs people looking for the product you are offering. These backers appreciate risk-takers' love to look at new products, and this is an excellent opportunity to reach them.

Crowdfunding allows you to present your product to an open community where you can begin to gauge its attraction to potential customers. You can receive helpful feedback and use the insights to improve your product. You can potentially validate your mission and then push the promotion over to your website—and, with your social media campaign, drive more interested people to your product. Your crowdfunding campaign can become your worldwide marketing platform.

Crowdfunding Disadvantages

Development of your marketing campaign is vital to driving awareness, sales, and getting off the ground. If you are not experienced in marketing, this will give you much-needed experience, but it may also cost you some serious money. You will need robust marketing on social media sites such as Facebook, Instagram, Twitter, Pinterest, Tumblr, and LinkedIn. There is now a cottage industry with full service companies springing up to help launch your campaign, who can help with prototyping, filming, and social media marketing.

With every crowdfunding platform, there is an immense amount of fine print. I encourage you to read and understand it all. There is always language explaining that crowdfunding sites are not responsible for any product wrongdoings, and what you do to satisfy orders is your responsibility. You are also the party that is solely responsible for consumer satisfaction. Of course, this helps to protect them if you don't come through, but it will lead to a lot of problems if you shut down without fulfilling all the orders.

Using Crowdfunding to License a Product

I think it's brilliant when inventors develop a strategy to build crowdfunding awareness of their product and use it to leverage a great licensing deal. I've put together many licensing agreements from Kickstarter and Indiegogo campaigns. I comb the sites every day. The exposure these platforms provide is unprecedented. I have entered some licensing contracts where the company takes over manufacturing the product, and the inventor begins to collect royalties immediately. You may initially want to go directly to market entirely on your own, but when you find out how much work it takes, change over to licensing. Once again, it's a brave new world out there today. The good news, though, is crowdfunding has been one of the more pleasant recent surprises.

16

Direct Response Television

DRTV or Direct Response Television ads, often referred to as infomercials, are where TV and internet viewers may order an advertised product directly, using a toll-free number or visiting a website landing page. It's an impulse buy that works best when a problem-solving product is demoed effectively and sold at the right price point. Ordering is usually effortless, and the product typically arrives within a few days. What could be easier? The operators standing by are trained to add to the customer's initial order by encouraging multiple sales and add-ons, something retailers have been promoting since the beginning of time. With a compelling sales pitch, a $25 order can often be turned into a $70 or $80 purchase. It's a big business.

I was fortunate to be involved, a few years ago, launching via TV infomercial the hit product Smart Spin. The product was a $19.99 kitchen cabinet shelf storage container and lid organizer invented by Boston-based Saul Palder and rolled out by DRTV Guru and

Merchant Media owner Michael Antino, in conjunction with All-Star Marketing. Almost overnight, Smart Spin became a household brand. Within the first year, we sold over five million units, and I learned firsthand the enormous power of sustained television marketing. Not all products are suitable for DRTV. The odds are long in finding the right ones. Nevertheless, when a product hits and rolls out with TV support ads running and retail store placement, the sales can be significant.

DRTV and Licensing

I enjoy explaining to inventors the differences between licensing with a traditional manufacturing company and with DRTV companies. The two approaches and cultures are quite different. Traditional manufacturing involves exploiting brands and leveraging retail shelf space through good old-fashioned sales connections and consumer loyalty. These products can often stay on the shelf for a long time. TV commercials, on the other hand, can dramatically speed up the consumer awareness process and put high octane fuel into a product's launch. For inventors, DRTV is a way to get your product in front of millions of consumers almost overnight aided by a visual demonstration of how the product works and its benefits. On the other hand, the sales run is usually much shorter, typically a year or two, and then it's over. It's a shorter, steeper curve.

DRTV is by no means for every product or person. The window for successful products is small, and competition can be fierce. Unless the product can sell a minimum of a million units, it probably will not fit into the high-volume DRTV model, so get used to rejection. The product must have mass appeal. In other words, the problem you are solving must be shared by enough people to generate tremendous appeal. It also requires a terrific video demonstration.

Because of its potential explosiveness, there is a lot of appeal in this category for inventors. In general, DRTV products solve a real-world problem in unique ways at an affordable retail price. A successful product pitch compels the customer to get off the couch, go to a phone or computer, and order right away. Though the commercials may run late at night and even look a little cheesy, it's a very analytical industry run by some brilliant people. These experts have reviewed many products and conduct a great deal of sophisticated consumer testing behind the scenes before moving forward with a rollout. A rollout can cost millions of dollars upfront to produce sufficient inventory and run a significant number of ads on many stations in many TV markets. They do not spend their money foolishly. Once the commercials take off, the product is also rolled out into As Seen On TV areas in all the major retailers across America. You will find these products on well-displayed endcaps that can't be missed because once the product has been seen enough times on TV, most are sold at retail. The suddenly overnight brand sensation is recognized in stores and added to the shopping cart.

INVENTOR INDUSTRY ROCKSTAR

Mike Lindell

Mike is best known as the inventor and CEO of MyPillow, a product he invented and business he grew into a major Minnesota manufacturing company. Mike also has sincere enthusiasm and heart for helping people recover from addictions, which inspired him to launch the Lindell Recovery Network. One of the things that makes Mike's story so great is that he broke into the DRTV business as an outsider and has now sold fifty million pillows in less than ten years! Mike employs well over a thousand people here in the United States. His approach to analytical data collection when TV and radio ads are run is highly sophisticated and enables tweaking of offers and special focus placed on consumers by specializing geographic and demographic reach. After a variety of life challenges, Mike became an "overnight" sensation, which no doubt took a great deal of determination and hard work. As a person and company, it's what this country is all about.

The DRTV Industry Continues to Change

DRTV started back in the 1960s with Ron Popeil and his Pocket Fisherman, which was the natural extension of ocean boardwalk pitchmen back in the 1920s—which was the direct descendant of the original covered wagon pitches back in the 1800s. Everyone can remember their favorite commercials like the Ginsu knives, yet, the industry drastically changed when products from the commercials became available in retail stores, a much bigger marketplace.

Today, internet sales and other electronic media drive the business. Traditional TV viewing has decreased with the introduction of livestreaming, plus the consumer's ability to record television programs and skip commercials. The DRTV industry has recognized this shift and adapted to other electronic and digital platforms. Today, consumers may shop for products everywhere, including their mobile phones, tablets, electronic devices, and laptops. They can also buy products from online ads with the click of a button without sitting through an entire commercial. Companies are also tapping into focused demographics and markets through artificial intelligence. Even though DRTV remains a recognizable acronym for the industry, the category has now expanded to include many high-tech arenas.

DRTV Formats (Long and Short Form)

We have all been up late at night, turned on the television, and been confronted with "Call in the next ten minutes, and we will double your offer!" These commercials are both appealing to many consumers and strategically placed. Once you see the same commercial five to seven times on TV, brand awareness rises. Statistics show that only one out of twenty infomercials have a successful product rollout. So, what may look easy or lucky is anything but, which is why it pays to work with companies who are historically successful in this space.

There are two forms of DRTV commercials, short-form and long-form. Short-form is those fast-paced, hard-sell commercials running between one and two minutes and primarily featuring products in the $19.99 to $39.99 range. Long-form infomercials last approximately thirty to sixty minutes and feel like they were shot on a movie set. These commercials target a higher income demographic, selling products from $99.99 to $499.99, sometimes offering multiple payment options. Either way, both are intended to sell enough units to cover the cost of production and rollouts, and build instant brand recognition.

Television Shopping Channels

Don't confuse the DRTV business with home shopping channels such as QVC or HSN. Though products are sold through these venues using television as the primary medium, these businesses are more like traditional brick-and-mortar retailers, ordering merchandise from consumer goods manufacturers, adding their margin to the mix, and warehousing the necessary inventory to support sales. DRTV, on the other hand, markets to the consumer directly, generating pent-up demand. At some point, when enough demand has been created, you will see the DRTV product on QVC or HSN, just as you will see it on the shelf at Target or Walmart.

INVENTOR INDUSTRY ROCKSTAR

QVC On-Air Personality Scott Hynd

Scott is the president and co-founder of Proformance Marketing Group (PMG), a Philadelphia-based marketing firm that specializes in Direct Response Television and product promotion. Scott appears on-air regularly as a product expert both on QVC and in infomercials. Over the past decade, Scott has presented over 250 different products on national television, totaling thousands of appearances and well over a billion dollars in sales. With a passion for sharing his knowledge and expertise, Scott is a prolific educational speaker, panelist, and judge at industry trade shows and inventor workshops across the country. His presentation on how to pitch a product is priceless.

Stringent Product Testing

DRTV companies don't just stumble upon products. The consumer testing process is analytically based on sales data and historical success matrixes. Companies that have had winners in the past are better suited to predict winners in the future because they have all the sales data on hand. Testing often begins with simple consumer surveys sent to random people determined by demographic. They are asked basic questions about product benefits and their likelihood to purchase at an established price. The results of these surveys are measured against other successful products from the past. If the product clears the first test hurdle, it usually moves onto a much more elaborate web test that goes to many more people. Sometimes they will even ask for a credit card to verify the interest is genuine, although the transaction never processes until the goods are ready for delivery. The third wave of testing, if the product survives the first two, is a real on-air TV ad that costs a fair amount to produce and runs in many markets across the country. The sales results here are critical. The ads need to drive enough sales to sustain the cost of the TV commercial campaign. Even though this may seem like a considerable amount of testing, it's all done quietly under the radar and is never noticed by the vast majority of American consumers.

Preparing for DRTV

If you are planning to pursue a DRTV licensing agreement, you should adjust your product presentation to reflect the DRTV marketplace. Here are a few things to consider when approaching DRTV companies.

Sell Sheet

Put together a sell sheet with a professional image of the actual product and a benefit statement, including a list of significant consumer benefits. Make sure it's only one page, save it as a PDF and place a direct link to your product demonstration video on the sell sheet.

Demonstration Video

This is extremely helpful in this arena because it's a preview of how the ad might look and feel on television. Your video should be concise and last no longer than one minute. Show your product in action in a DRTV-style commercial format, if you can, by exposing the problem and how the product solves the problem and product benefits. The video does not have to be professionally made; however, it should be scripted out and kept simple.

Prototype

Make sure you have a functioning prototype to shoot a persuasive video. DRTV companies tend to be a little easier to approach than traditional manufacturers if your prototype is not perfect. If they like the concept, they might be able to help move things along. And remember, the more prepared you are, the higher the royalty rate you can command.

Intellectual Property Protection

As with prototypes, DRTV companies are also a little less demanding on the patent front. Because they want to get the product to market very quickly, sometimes they will put less emphasis on intellectual property in the evaluation. Remember, a lack of patent protection will lead to a lower royalty offer, as well as open the product up to being knocked-off in a highly competitive industry, so you should keep your inventing standards up. I would still encourage utility protection. Registering a trade name is usually not crucial because DRTV companies will test multiple names to see which works best and select one they can eventually own. Since they are paying for all the advertising, you can hold the patent, but they want to own the trademark.

SECTION FOUR

Organic Innovation Matters

17

The Importance of Organic Innovation

I consider myself more of a biologist than a chemist. I love organic growth. Planting a seed and watching it bloom is for me. All-natural. Chemists are more likely to pursue predictable formulas that result in more predictable outcomes.

On the other hand, biologists nurture life. Which do you think most inventors are like, biologists or chemists? Are invention outcomes predictable? I think you know where I am going with this. Every inventor needs to be, at least in part, a biologist.

In the inventing world, I don't think we can always predict natural growth and success. An inventor needs to be patient, persistent, dedicated, and attentive, just like farmers growing their crops. True inventing is embarking on a journey with an unknown destination, every product a unique ride. It's essential to monitor on-the-ground conditions as the weather changes, and you gather more information. A remarkably successful inventor once told me that every morning they arise with a list of challenges to address, get as much solved during the day as possible, and then start the entire process over the next day. On one project, they did this every day for over a year until all the problems were solved.

The Glass House

In my hometown in Connecticut, there is the famous Glass House, designed by Philip Johnson, a world-renowned architect. He considered himself more than an architect. He was also a landscape artist. Although he designed the physical structures on the compound, he also planted all the trees in different locales throughout the property. He had the unique vision to know where best to plant, affecting both the property's function and aesthetics in the future. This hit me as the ultimate in confidence and patience. As an inventor, you must be able to foresee your product in full bloom, being both flexible and steadfast, parts patient and visionary, believing and understanding where your product needs to reach consumers in advance. Philip Johnson was both organic and visionary.

Inventing Involves Real Work

Few significant accomplishments worth pursuing are easy. Anyone who says differently is trying to sell you something. Inventing isn't easy either. It takes a great deal of hard work, tenacity, endurance, intelligence, and stamina. Ask any experienced inventor, and they will tell you that you can even learn from your failures. Thomas Edison was quoted in saying, "I have not failed. I've just found 10,000 ways that don't work." Failure and mistakes are a natural course in finding the ways your product will work. If you work hard, your own product development process will guide you to take the right direction.

Inventing is about controlling your emotions, overcoming adversity, and staying true to yourself. In the beginning, there is a natural beauty being in the throes of trial and error, doing things the right way with heart and soul, pushing yourself through the obstacles. The key is to stay motivated, not take shortcuts, and push yourself to find unique and functional solutions to real consumer challenges.

One big obstacle for inventors is getting on, and staying on, the right track. How do you even know when you are also on the right track? Without a compass, it is easy to get lost. Losing sight, direction, and what you want to do with your invention puts you in a precarious position causing you to seek help in the wrong places. This is the stage where you must keep your head about you, and in this vulnerable place, watch out for those who would take advantage.

INVENTOR INDUSTRY ROCKSTAR

Chicago Inventors Club Calvin Flowers

Calvin is the founder and current executive director of one of the best local inventor clubs in the country, the Chicago Inventors Organization. Responsible for implementing club policy and overseeing the day-to-day operations, Calvin has grown the club from scratch to well over 2,000 members. Calvin, an inventor himself, brings in regular speakers monthly to share educational information and runs an annual all-day conference with experts attending from around the country in all fields that assist inventors, including patent attorneys, prototypers, and USPTO officials. What he does every day to help others, out of the goodness of his heart, is inspiring.

Feel Better About Your Product Every Day . . . Or Move On

Most inventors carry around countless ideas in their heads, many of which will never see the light of day. Yet, there is always that special one that stands out. You know it's great because you wake up every morning and the idea feels better than the day before. Then, you start putting it through its paces, trying to get it on the right track, taking an exciting core idea and pushing it to be successful. You begin the process of validating, testing it in your mind against everything going on around you in the world, and one step at a time it clears hurdles that you know are ahead of you. And, of course, you proceed with the development steps that we've talked about throughout this book. Life is good.

At the same time, and this is important, if you wake up one morning and the idea no longer excites you, it's genuinely not a good sign. If you have the same negative feelings several days in a row, or at any point in the development process, you need to consider moving on to another idea. Organically, it sometimes is equally important to let the project die, to let go and move onto something else without spending a lot of energy and money. You need to allow nature to take its course.

In reviewing new products every day, the exact same process happens with me. On average, I receive over twenty outside product submissions a day. Of those, half I can instantly tell are not ready. Either the product has been done before, or is outside my arena of expertise, or it's simply not ready for prime time. Of the remaining, there may be a few that genuinely spark my interest. Then there are typically a few others that I feel may have some merit, but I'm more on the fence about. These are always the toughest ones to decide upon. I've learned to place them to the side and come back in a day or two and review them in a new light. If I am still not excited, I know the answer in my heart and move on. Every so often, I get a better feel for the product, and my enthusiasm grows. Either way, I let the review process play out organically, so I don't miss anything important. It's advice I would also provide inventors. Slow down and don't force things. Let them grow. The very definition of organic is a natural process. A good product idea must evolve naturally through the right incubation process and, eventually, development protocol. Product ideas have a life of their own, and as an inventor, your job is to foster it—or let it go.

Passive Income and Feudal Lords

Organic growth and inspired innovation are not about generating passive income, defined as revenue that requires little effort to maintain. I mention this because many inventors have been told they can license a product idea, sit back, and passively collect an income. This is poor advice. If this ever happens the way it's marketed to inventors, it is extremely rare.

I think the passive income gig must have started back in feudal Europe in the dark ages, when peasants and serfs labored the land, while the landowners sat in their castles, overseeing the workers below and made a fortune off others. Ever since, some in society seem to aspire to work less and making more, even while others suffer. If you think you're going to make money doing nothing in the invention world, good luck. Sorry, but getting rich these days isn't for sissies.

Profitable inventing takes time, effort, and heart. There is no substitute for hands-on developmental work, tilling the land, and watching the growth every day. Shortcuts simply don't work, just like growing organic food. When you use ingenuity and develop something unique, there is a special feeling and understanding of yourself that results, an inner pride that you didn't steal someone else's idea to further your own interests. Like planting a tree, the beauty comes later, sometimes years later.

New products often fail. It's the cycle of life in this business. You will not be alone, as many famous inventors throughout history were risk-takers and failed. Tesla failed with his *Thought Camera* and Thomas Edison with his *Mass-Produced Concrete Building*. However, each of them moved on to other concepts and kept pushing the envelope. Solving problems is the foundation of great inventing, and following the slow and steady organic development path is your ticket to profitability.

Innovating Our Way Out of Recession

As of this writing, in the spring of 2020, the country, and the world for that matter, is awash in the COVID-19 pandemic. Being quarantined, like so many others with their families for months, I have been thinking about the essential role innovation can play during these extremely difficult days—and its role ahead in digging out.

The pandemic appeared seemingly out of nowhere and has locked our country down. Public events have stopped, and mandates are in place to remain inside, wear masks, and prevent the spread of the virus. Yet, at some point, people will re-emerge, and both life and our economy will slowly start getting back to normal. How will innovation organically grow out of this crisis? How can it help us recover? How will the future play out in both the short and long terms?

During the crisis, everyone in America was in the same boat, forced to protect themselves and families and stay at home, as the sickness prompted a need for supplies and protection devices, garments, and equipment. There has been an amazing push across

America to address many needs head-on. Thank God for our first responders. Many people and companies are racing against the clock to develop therapeutics, vaccines, and technologies in order to save lives. The American people require safety, and addressing such life and death challenges calls for innovative solutions.

On a side note, it will be interesting to see if the traditional anti-patent groups and companies use this unusual time to claim any intellectual property for life-saving devices and medicines should not be patentable. That position is shortsighted in many ways, but may prove popular in the media. I truly hope that one of the unintended consequences of the pandemic is not a further weakening of our patent system, at a fragile time in our history, championed by those who do not understand the power of financial incentive and raising startup capital.

Like other challenges, COVID-19 has also had some unforeseen side benefits. First, life has slowed down dramatically. Most have had time to reflect and take stock of what's truly important. I, for one, had the opportunity to finish this book. The air around us has quickly become cleaner with reduced emissions, in ways softening the footprint of man on our planet. The economy has shifted overnight with less local and global travel, which has reduced energy consumption, which has possibly created the perfect time, with all of our environmental concerns, to start shifting our habits and developing new solutions in this important arena. It will be interesting to see if we gain momentum as a society during this time-out to clean up the planet, or we simply go back to business as usual and unabashed consumption when the crisis is over. We'll surely know by the time this book is published. I, for one, am hoping inventors staying at home, with extra time and seeing the benefit of clean air around them, will double down and make new strides and develop profitable inventions to make all of our lives healthier and better.

This leads me to think about an article I once read about a person who was a raving alcoholic and hit by a car. He was bedridden and in a coma, stuck in the hospital for six weeks. Eventually, the person recuperated, and when he finally left the hospital, as an unexpected side benefit, his blood was entirely cleaned up. At least temporarily the toxicity of alcohol was cleared from his body. I am wondering if that same cure and side benefit might happen to us with the environment. The sad news, the article ended with the alcoholic resuming his drinking ways. He passed away only two years later. He could not help himself. Let's see how we do. Hopefully, a lot better.

Great innovators always seem able to adapt culturally to the abrupt changes our society presents. Many things will change dramatically after the coronavirus. There will obviously be a need for a vaccine and better medical equipment. There will also likely be a need to continue social distancing. Who knows, maybe we will stop shaking hands altogether. Even social gatherings, sporting events, and life events such as marriages, graduations, and childbirth rituals may be dramatically altered.

Yet, at the same time, will the door be opened to tremendous new opportunities on the innovation front? People have started working more from home, and school classrooms are being held virtually. Will this trend continue and change forever the way we work, learn, and gather? Companies have quickly adapted, establishing video networking to pursue business, and continue meeting virtually. Inventors who come up with better solutions in this arena should prosper.

Who will come out as financial winners developing the new technologies needed to facilitate this shift in our culture and ways? Be visionary. It could be you. Now that you know how the game really works, don't just mail it in; be sure to pursue innovation the right way. Our country is counting on you. As our economy heals and investment begins flowing again, it is imperative to have a strong patent system that protects and rewards the necessary risk and disruptive solutions we, as a society, demand. Grassroot inventors should be supported and feel fully vested in the nation's problems and solutions. Innovation will eventually lead us out of this economic recession, and, if history repeats, inventors will play a big role.

18

The Future of Innovation in America

Everyone Can Do More to Help

Assuming you've made it this far and absorbed the primary themes of the book, I'll keep this chapter short and sweet. Quite simply, here's what we can all do to ensure that America remains the most innovative place on the planet, while restoring pride and purpose to our natural disrupters.

Innovation is an overused marketing buzzword that's tossed aimlessly about these days. It's a term that feels as American as apple pie, however most of our country does not realize that behind the scenes, the state of American innovation is at a serious crossroads. If we don't begin collectively playing our cards right, innovation will no longer be a worldwide American-led standard. It will move offshore to other countries, just like the auto industry did back in the 1980s. Were you aware, for instance, that more patents are now being filed in China than the United States? Who would have thought this only a few years ago? We need to return the profit incentive, and for inventors to once again challenge the American status quo, otherwise, we will wake up one day soon and realize our generation blew yet another good thing.

I have written throughout this book about many myopic forms of self-interest sweeping through America's innovation space and its debilitating effects. Inventors should be aware of the changing landscape and recognize the significant challenges before charging ahead and wasting your hard-earned savings on a dream without rational foundation or prospect. I'll end with what inventors can do, but allow me first to describe what each separate player in the American innovation sandbox can do to improve the current state of affairs.

USPTO

The United States Patent and Trademark Office should bring transparency and better understanding to the entire Patent Trial and Appeal Board (PTAB) process, select judges that have the necessary credentials and experience without personal bias or agenda, and generate more trust with inventor shareholders. In order to raise capital and grow small businesses, only a small percentage of patents should be rejected after they have been examined and issued. Know that our entire economic system relies on reliable information that should not crumble under the political pressure generated by wealthy big-tech companies. And please embrace your pivotal role in providing a fair-minded foundation for advancing innovation in America. It is, after all, the USPTO's essential calling.

Congress

Pass legislation that corrects the poison-pill validity issues within the PTAB portion of the AIA, which was cleverly written by big-tech company lobbyists. Everyone in Congress should do a better job of understanding the tremendous positive effect that disruptive innovation has on America's economy and our quality of life. Encourage independent innovators in the next generation by increasing every American's fair and profitable access to our innovation ecosystem. And listen carefully to your members who are inventors themselves.

Courts

Work harder to understand the technically complicated, patent-related cases in front of you. They may not be particularly topical in the news, but your decisions matter. And, to Clarence Thomas and the Supreme Court, patents are not a franchise; they are a property right! Please go back and look at the Constitution, Article 1, Section 8. Let's begin rebooting our current skewed intellectual property thinking and rediscover what our founding fathers intended.

Manufacturing Companies

Be more open to receiving outside product submissions, initiating licensing agreements, and bridging the culture gap with the independent inventor community. In the long run, this will be far more profitable to you, inventors, and society, than the short-sighted pursuit of stealing or working around other people's intellectual property. Progressive open innovation programs can work.

Big-Tech Companies

Please, it's not *all about you*. Think about your personal legacies. Understand that when you promote onerous legislation for your own self-interest, it has a debilitating side effect on the entire innovation ecosystem and can ruin people's lives. We can live together. Start by setting up great open innovation programs and paying fair and affordable royalties. There's an easy way to get this done; use the Singer

Sewing Machine royalty model. Look up the story, or call me, and I'll show you how to make it work for everyone. Don't forget, only a few years ago, you were the ones trying to launch your innovative platforms and were aided by a patent system that you are now, with newfound market dominance, diluting.

US Retailers

Care about where the products you sell actually originate, and I do not just mean geographically. Say no to obvious knock-off products that infringe on legitimate US patents, or are pirated. You are the primary gateway to most US consumers. Your ethics and actions matter.

Inventor Marketing Companies

Provide better services for what you charge and do the right thing by setting up a system where you either professionally counsel inventors to save their money by bailing out on unmarketable ideas or start collecting on the back-end of success. Please post your clients' real royalty success rates as a percentage of your total business sales. Additionally, for the inventor coach subset within this arena, stop being duplicitous about suggesting wealth, riches, and goldmines to those most financially susceptible and fragile in the inventor community. Try taking a different approach by "earning" your money partnering instead of charging non-refundable upfront fees. And, finally, when marketing, please differentiate between licensing deals that lead nowhere and real marketplace sales success, as well as better identifying the people you promote, as opposed to those you had no part in coaching.

Makers

Understand from some of your innovation predecessors that if you don't protect your intellectual property, it may get ripped off. Stop giving away your proprietary knowledge to larger communities that don't care about you as an individual. Otherwise, keep up the great work and keep doing what you're doing.

US Consumers

Give extra consideration, as you might to environmentally sound products, or locally grown food, or healthy living alternatives, to how a product was actually created, exactly like backers do on Kickstarter and Indiegogo. Recognize, before it's too late, that if we expect all wonderful, problem solving, creative new products will somehow, someday be developed only by large companies, we are in big trouble. Let's make it cool to be an inventor again.

Open Innovation Directors

Work harder. Keep spreading the word about connecting companies with outliers. Never stop giving back to others, always tell the truth, and get big companies to act ethically while believing in and promoting the benefits of the open innovation mission.

Inventors

Stop, look around, listen, and learn. Follow your intuition, make a plan, stick to it until rational feedback tells you it's not working, establish and maintain high product development standards, and don't get too far over the front of your financial surfboard. Please don't blow your money on empty promises, pretty faces, and inventor cults. If things don't work out the first time, save your shekels, get up off the mat, and move ahead as a much smarter person to your next project. Then, give back to the inventor community by sharing what you learned with others, without charging.

Peace.

Special Acknowledgment

I'm appreciative of all the folks at Lifetime Brands for their support over the years, but I'd like to thank three people in particular.

Today, Dan Siegel is the president of Lifetime Brands. When we started the Open Innovation Program in 2008, Dan was VP of Innovation. (I like to remind him that somewhere along the way I helped get him promoted.) Dan was integral to all the early success we had. Whether it was his background in housewares product sales or his love of good food, he has a terrific eye for new products. He can separate the clever from the saleable. Dan believed in the concept of reaching outside the company from the start, and we spent a lot of time together early on getting everything off the ground. He was vital to making certain the divisions were on the same wavelength and following through on projects internally.

Bill Lazaroff, the Executive VP of Product Design and Development at Lifetime Brands with about fifty developers and designers directly reporting to him, is responsible for launching thousands of new product SKUs for the company every year. Bill has a very similar retail background to mine, so we often see new products the same way and communicate together very well. Bill loves new products, and he is very smart and open-minded about innovation. He has always treated me as a friend and ally, not an outsider. This is critical to any successful outside contractor oriented open innovation program. I would like to thank Bill for introducing me to the SCAMPER and TRIZ ideation and development platforms.

I would also like to thank Lifetime Brands corporate counsel Sara Schindell for all her significant efforts since day one. Sara does the work of three attorneys and is always very fair but, at the same time, protective of the company. When we work on licensing agreements, she understands rational requests from the outside. She is also well versed on intellectual property, licensing, and contract law. Nothing gets by her. I worked closely in the early years with Sara to set up all the legal program guidelines to ensure that, along with her director of licensing, Terry Losco, everything from executing NDAs to licensing contracts moved forward in a timely manner.

I would also like to give a special shout-out to my friend Jeff Mangus. He patiently urged me for years to tell my unique story, at a time when I was simply too busy living it. He eventually wore me down with the persuasive argument that our industry needed to hear a different, non-marketer, non-sales, open innovation insider viewpoint. I could go further and describe two years of details, logistics, reconciliation, and lots of hard work, but they would be mere words. I'd rather simply say, thank you kindly Jeff.

Sources

1. "Inventiveness and the NoPhone," Inventors Digest, December 4, 2018. Accessed at https://www.inventorsdigest.com/articles/inventiveness-and-the-nophone/.
2. Malcolm Gladwell, *Outliers: The Story of Success* (New York: Little, Brown, and Company, 2008).
3. "Article I," Cornell Law School, Legal Information Institute. Accessed at https://www.law.cornell.edu/constitution/articlei.
4. "Appendix R: Patent Rules," Title 37—Code of Federal Regulations: Patents, Trademarks, and Copyrights. Accessed at https://www.uspto.gov/web/offices/pac/mpep/consolidated_rules.pdf.

Index

About the Author

WARREN TUTTLE has for many years overseen the Open Innovation product programs for several industry-leading companies, including Lifetime Brands in the housewares and tabletop arenas (Farberware, Kitchen Aid, and forty other brands), Techtronics Industries Power Tool Group in the power tool and hardware industries (Ryobi, Rigid, Hart Tools, and other brands), and Merchant Media in the Direct Response Television category (Smart Spin, True Touch, and many other brands). He was also behind the launch of several highly successful consumer products, including Misto, The Gourmet Olive Oil Sprayer, and the SmartSpin Storage Container System. Warren personally interacts with many thousands of inventors every year and has initiated well over one hundred new consumer-product licensing agreements that have collectively generated over a billion dollars in retail sales.

Warren is also a well-known advocate for inventor rights. He serves as the president of the United Inventors Association Board of Directors, a national 501(c)(3) non-profit with high ethical standards that helps inventors through education, advocacy, and the sponsorship of inventor booth pavilions at several industry trade shows, most notably the National Hardware Show. Warren is also a member of the National Pro Bono Patent Commission and the US Patent and Trademark Office National Council for Expanding American Innovation.

He lives with his wife, Lynn, in Southern Connecticut and has three wonderful daughters, all well-educated and working. He enjoys skiing, motorcycling, golf, and travel. Please visit Warren's website, www.tuttleinnovation.com, or email him at wwtuttle@yahoo.com for more information.